You and your genes - Higher

1 What makes you the way you are?

a Complete the sentences. Use words from the list.

genes	alike	variation	environment	unique	features

Human beings are all very __alike__ . We share many of the same __features__ . But

each person is __unique__ . There are differences between our features. We call these

differences __variation__ .

Differences between us are caused by:

➔ __environment__

➔ __genes | heredity__

➔ a mixture of both

b Look at the diagrams. Write down examples of variation under each heading.

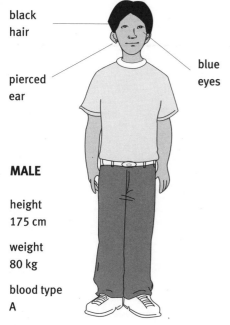

black hair

pierced ear

blue eyes

MALE

height 175 cm

weight 80 kg

blood type A

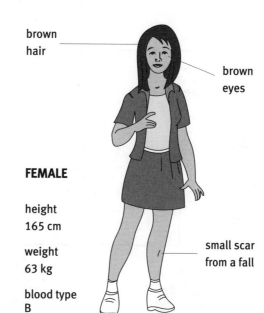

brown hair

brown eyes

FEMALE

height 165 cm

weight 63 kg

blood type B

small scar from a fall

Environment	**Genes**	**Both environment and genes**
Small scar	Blue Eyes	height
Pierced ear	Brown Eyes	Weight
	Brown Hair	
	Black Hair	
	Blood Type	

3

2 Cells, genes, and proteins

a Complete the missing labels by using words from the list.

cell	DNA	protein

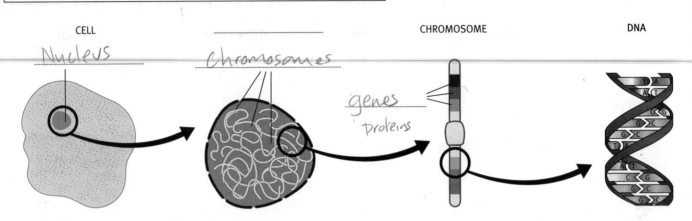

CELL · Nucleus

Chromosomes

genes
proteins

CHROMOSOME

DNA

b There are two types of protein. Describe their jobs in a cell.

CHECK

- Structural proteins: ?? body building proteins

- Enzymes: Speed up chemical reactions in the body
  eg in respiration, digestion

3 Family trees

This diagram shows a family tree.

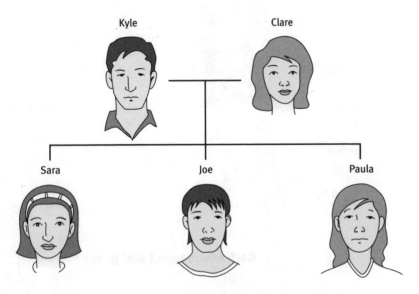

Kyle   Clare

Sara   Joe   Paula

a Explain why members of a family may look like each other.

Because the each get 23 chromosomes from their mother + father.
They share the same genes.

b A family tree helps to show how a feature has been passed on in a family.

Suggest one feature that
➔ Sara has inherited from Kyle: *Shape of nose*
➔ Sara has not inherited from anyone: *hair style or length of hair*
  (it is NOT affected by her genes)

---

4 One gene or more?

Most of a plant's or animal's features are controlled by more than one gene.
Only a few features are controlled by just one gene.

Look at the graph of variation in two features.

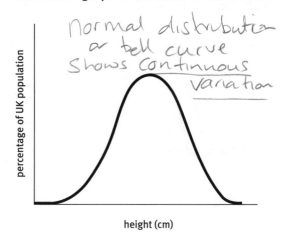

Normal distribution or bell curve shows continuous variation

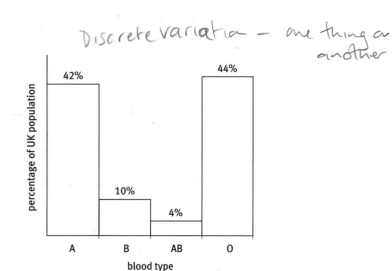

Discrete variation — one thing or another

a Which feature is controlled by:

just one gene: *blood type*   many genes: *height*

b Explain why you think this.   *check*

Height varies a lot. Blood type has only 4 groups. you have to be one of these groups

c What else affects a person's height?

What you eat, the Environment, your lifestyle, where you live

5 Clones

The plants in this diagram are clones.

a Explain what is meant by a clone.

They are genetically identical

without sex

b Name two types of living things that reproduce asexually to form clones.

1 *Cells*
2 *Cuttings from plants*

c The two plants in the diagram are clones. But they are not identical.
Explain why the plants can be different from each other even though they are clones.

The environment is different - different soil,
different amount of water or sunlight,

d Animal clones are produced naturally and artificially. Write short notes around these
diagrams to explain how clones are being produced.

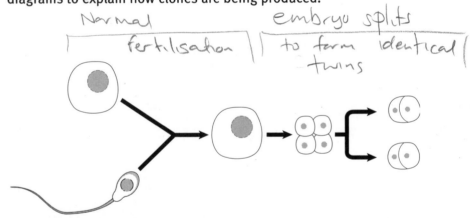

Normal fertilisation

embryo splits to form identical twins

Natural animal cloning

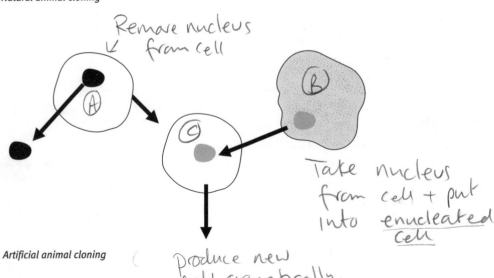

Remove nucleus from cell

Take nucleus from cell + put into enucleated cell

Artificial animal cloning

Produce new cell genetically identical to B

6   Using plant clones

Cells of multicellular plants and animals become specialized as the organism develops. Some plant cells remain unspecialized. They can develop into any type of plant cell. So they can re-grow parts of the plant if they are cut off. Growers use this to make plant clones by taking cuttings.

a Describe one method for making plant clones.

Cut cutting out of the plant. Put in soil, give it water eventually it will grow to the plant identical to the other plant.

b List at least one advantage and disadvantage of growing plants by cloning.

Advantages of cloning plants	Disadvantages of cloning plants
Same as the other plant	*No variation*
s	

7   Male or female

a Complete the sentences.

Human beings have ___*23*___ pairs of chromosomes in a normal body cell.

One of these pairs is called the ___*Sex*___ chromosomes. This pair controls

whether the person is male or female.

In a man the chromosomes in this pair are different sizes.

There is one ___*X*___ and one ___*y*___ chromosome. The pair is called ___*Xy*___.

In a woman the chromosomes in this pair are both the same size.

There are two ___*X*___ chromosomes. The pair is called ___*XX*___.

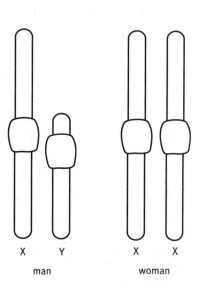

X   Y
man

X   X
woman

b Complete the diagram to show how a person inherits their sex chromosomes.

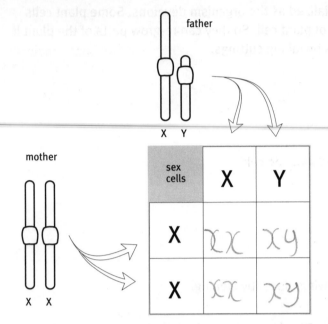

father

X   Y

The chance of any child being male is 50 %.

The chance of any child being female is 50 %.

mother

sex cells	X	Y
X	XX	XY
X	XX	XY

X   X

c A couple have two girls. They think it is more likely that their next child will be a boy. How would you explain to them that this is not the case?

there 50% change that it a boy and 50% that it is a girl. So your ~~Ne~~ Next baby could be a girl or a boy

d A gene on the Y chromosome controls whether a person is male. Write short notes in a flowchart to explain how this works.

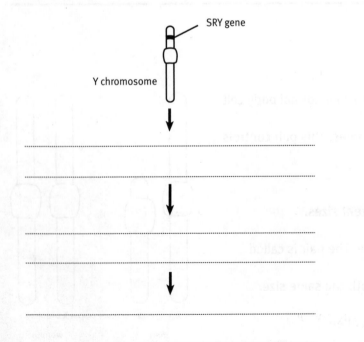

SRY gene

Y chromosome

(SRY stand for ....)

8   Fertilization

a Complete the sentences.

Sex cells carry copies of parents' chromosomes. At _fertilisation_ the nucleus from a male and

female sex cell join together. This gives the fertilized egg cells _4 6_ the number

of chromosomes.

The diagram shows a human sperm cell fertilizing a human egg cell.

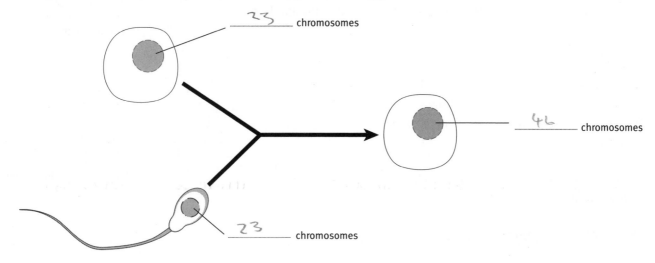

23 chromosomes

46 chromosomes

23 chromosomes

b Label the diagram to show how many chromosomes there are in the nucleus of each cell.

c Label the diagram to show whether the chromosomes are in pairs, or singles.

d These statements describe how chromosomes are passed on from parents to their children.
The statements are jumbled up.
Number the statements in order so they tell the story of fertilization.

1	Chromosomes in one of the mother's cells are copied.
☐	Each sperm cell nucleus has 23 single chromosomes.
☐	A new human baby develops.
☐	The sperm cell nucleus enters the egg cell.
☐	The fertilized egg cell nucleus has 23 pairs of chromosomes.
☐	The cells specialize into different types.
☐	Just one of the copies from each pair of chromosomes goes into a sperm cell.
☐	Each egg cell nucleus has 23 single chromosomes.
☐	The egg cell is fertilized.
2	Chromosomes in one of the father's cells are copied.
☐	The fertilized egg cell grows and divides many millions of times.
☐	Just one of the copies from each pair of chromosomes goes into an egg cell.

9   Genes are in pairs

a The diagram shows a pair of human chromosomes.
One of the genes they carry is marked on the diagram.

Complete the sentences.

Each human body cell has ___46___ of each chromosome.

Chromosomes of the same pair all carry the ___Same___ genes

in the same ___chromosomes___. This means that each human body

cell has ___copy___ of every gene.

b Sahira was born in Delhi, India. Pierre was born
in Paris, France. The diagrams show some of
the genes on both their chromosome pairs
number 7 and number 15.

Complete the chromosomes to show the position
of the labelled genes.

Use a different colour for each gene.

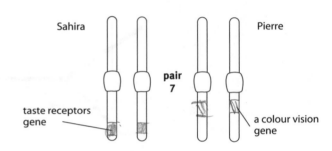

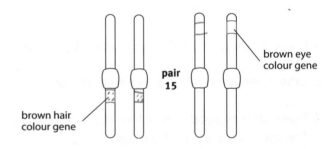

10 Alleles

Chromosome pairs carry the same genes. So genes also come in pairs. But pairs of genes are not always
identical. Genes may have different versions. Different versions of a gene are called **alleles**.

In humans, if you have dimples or not is controlled by one gene. This gene has two alleles – one for dimples
and one for no dimples. The allele for dimples is **dominant**.

a Explain what is meant by a **dominant** allele. _It will be shown even if there one dominant_
Allele;

b Explain what is meant by a **recessive** allele. _It need two recessive allele to be shown._

c Scientists use letters as shorthand symbols to write down alleles.

Write down the symbols for:

allele for dimples = _D_ (dominant)

allele for no dimples = _d_ (recessive)

d Some of the people in the family tree shown below have dimples.

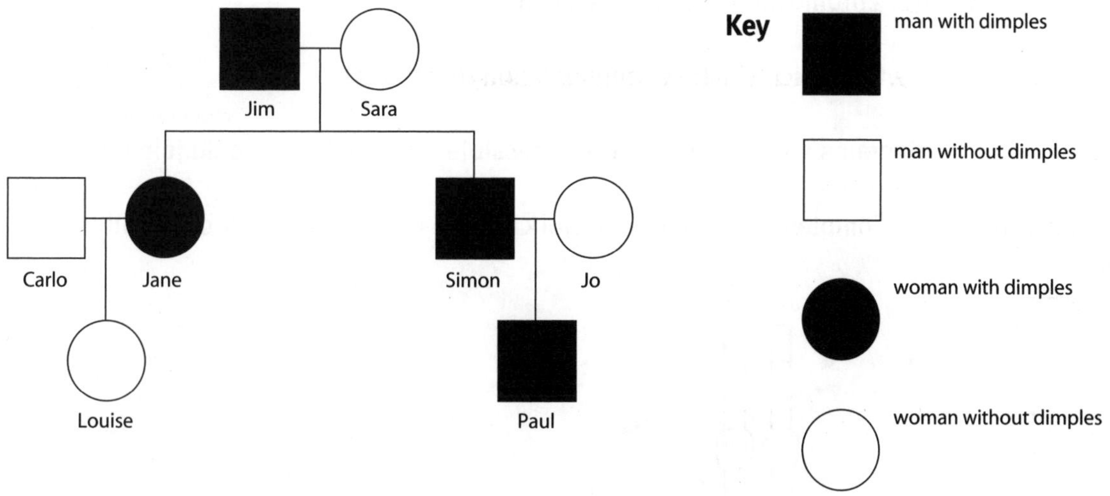

Key
- man with dimples
- man without dimples
- woman with dimples
- woman without dimples

i Name one member of the family who has dimples. _Jim_

ii Explain why Louise does not have dimples.

Because She has two recessive Allele

---

11 Huntington's disorder

Huntington's disorder is an inherited disorder. It is caused by a dominant allele.

Describe the symptoms of Huntington's disorder.

Memory Loss; can't control the body.

12 Genetic crosses

a Complete the sentences. Use words and letters from the list.

Y	one	cross	half	X	fertilization

Sex cells in humans are ____XX____ and ____XY____. Every sex cell carries a copy of

____Half____ the parents' chromosomes. Each sex cell has only ____One____ of each

chromosome. It is not possible to predict which sex cells will join together at ____Fertilization____.

Genetic ____Cross____ diagrams show all the different possible ways that they could join up.

b A couple are expecting a baby. Complete the diagram to work out the chance that their child will have dimples.

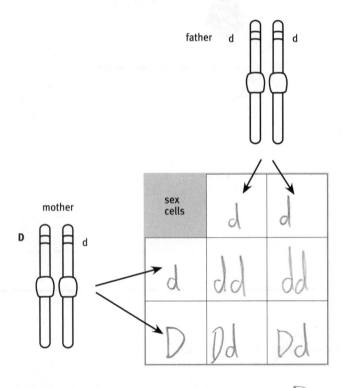

father   d   d

mother

D   d

sex cells	d	d
d	dd	dd
D	Dd	Dd

Key
D = allele for dimples
d = allele for no dimples

The chance of the baby having dimples is ____50____ %.

c Explain why brothers and sisters with the same biological parents do not look exactly the same.

because they may not inheritied the same features that you can see clearly.

13 Cystic fibrosis

Complete the information on cystic fibrosis.

Symptoms:

mucas in the Luongs

It block tubes that take enzymes from the pancreas to the gut

Treatments:

Physiotheraphy & helps to clear mucus from the lungs.

Suffers take tablets with the missing gut enzymesin.

Antibiotics treat the chest infections

The allele is **dominant/recessive**.

Some people are carriers of the cystic fibrosis allele. Explain what is meant by **carrier**.

they don't have the cystic fibrosis but There children could have Cystic fibrosis.

14 Ethics

Some decisions involve ethics.

a Explain what is meant by an 'ethical decision'.

You decide what is right or wrong. ✓

b Give an example from everyday life of a decision which

➔ does not involve ethics: feeding your self ✓

➔ does involve ethics: terminate a baby

15 Genetic testing

Doctors can test adults, fetuses, and embryos for alleles that cause some genetic disorders. People may use this information to make personal decisions, such as

→ whether or not to have children
→ whether or not to terminate a pregnancy

The tests are usually accurate, but a small number of results will be inaccurate. Explain what is meant by

→ a **false-positive** result: *the amniocetesis test was unclear, when the baby was born she was fine*

→ a **false-negative** result: *very hard to make a decision ... which was ... she was fine*

16 Decision making

People make ethical decisions in different ways. For example, they may decide that something

→ should not be done because it is unnatural or wrong
→ is the right decision because it is best for most people

Doctors can test a fetus for the alleles which cause some genetic disorders. People have different views about whether this should be allowed.

Suggest three different opinions which people may have about this issue:

1 *It is unnatural to check what baby has*

2 *I don't care if my baby has it, I will still look after it.*

3 *If I test for the diesease, I might terminate the baby, I don't want to be terminate the baby.* ✓

17 Who should know about your genes?

Genetic testing is used by individuals to give them information about themselves and their family. It could also be used by other groups.

a For each of these examples, say why they want information from genetic testing:

→ Genetic screening programmes: *to see if genetic programme works - population*
→ Insurance companies: *to see which people Are more likely to need insurance* ✓
→ Employers: *to see which employer carry the dielecrei* ✗

b Choose one of the examples in part a and describe how people may be worried about this use of genetic testing.

insurance companies, they may raise the priece for the insurance because they got the diesease or carry the disease. ✓

check your detail when answering question
strive to facts and not waffle.
Complete Q 19-21   B7   B 14/10

c Do you think genetic information should be used in these ways? Say why you think this.

No only the doctor should use this information, the other people don't need this information.

18 Embryo selection

Genetic tests can be used to select embryos before they are implanted in the womb. This is called pre-implantation genetic diagnosis (PGD). PGD is only allowed in very specific cases.

Use this space to explain how PGD happens.

the patient takes a fertility drug
She release several ova
The doctor collects the ova
→ The man sperms fertize the ova in a petri dish (v itrofertilasation)

embryo reach the eight cell stage)
one cell is removed from rereach !
↳ the cells are tested for the disease

Only embryos without the diesase allele are implanted in sally's uterus.

19 Gene therapy

a Complete the sentences.

Some faulty ___*genes alleles*___ cause disease. It may be possible to treat someone with a disease

by replacing the faulty allele with a ___*healthy normal*___ allele. This is called gene ___*therapy*___.

It has been used successfully to treat a small number of children with ___*Scid*___.

Trials on people with ___*cystic fibrosis*___ have not been so successful.

b The diagrams show the main steps in gene therapy. Explain what is happening at each stage.

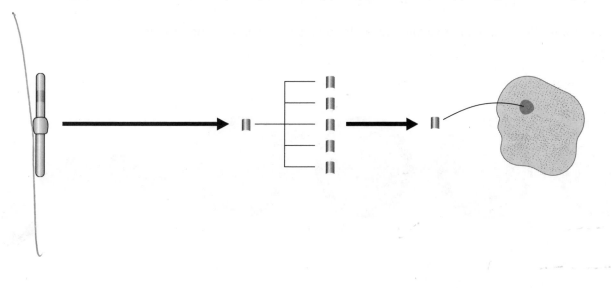

c People have different views about whether gene therapy should be allowed.
Suggest three different opinions which people may have about this issue:

1 ~~it does not ta~~ he effect does not last for long

2 ~~it that~~ It might have side effects

3 It is making my baby better, for exmaple Scid.

---

20 Stem cells

a Complete the notes about stem cells.

Stem cells are found in animals and plants. They are very important in the body because they are
unSpecialized so they could turn to any cell in the animal or plant

Doctors are interested in stem cells because they could cure some disease like
Parkingson!

Embryos and adults both have stem cells. Doctors think that embryonic stem cells could be more useful in

treating diseases because they are still not Specialized

When cells from one person are put into another person's body (in a transplant) they may be rejected.

In the future doctors could use embryonic stem cells from an embryo which has been cloned from

the patient's cells. These would not be rejected because it is from the same body and to the
Same blood group and genes

b Write notes around the diagram to explain how embryonic stem cells could be produced to
treat diseases.

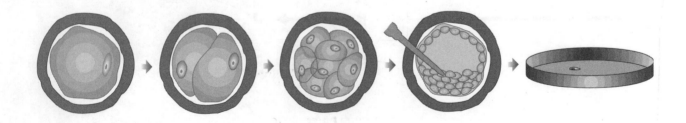

21 Cloning

This extract is from a newspaper report about cloning and stem cells.

Cloning for stem cells

What are stem cells?
Stem cells are unspecialized cells. Human stem cells can develop into over 300 different kinds of cells.

Where are stem cells found?
Stem cells are found in embryos which are just a few days old. They are also found in children and adults. For example, we all have blood stem cells which make new red and white blood cells during our lives.

Adult stem cells can only develop into one type of cell. But embryonic stem cells can develop into any type of cell in the human body. Scientists think embryonic stem cells will be more useful in research to cure diseases.

What are stem cells used for?
Scientists think that stem cells could be used to make new cells for people with some diseases, such as Alzheimer's, Parkinson's and diabetes.

The Royal Society supports research into the use of stem cells: 'Degenerative diseases and serious injuries to organs and tissues may be treated through stem cell therapies. Research on human embryonic stem cells will be required to investigate all of the potential therapies because other cell types, such as adult stem cells, may not have the same breadth of applications.'

What's this got to do with cloning?
Doctors already treat people with new cells when they do transplants. But when the new cells come from another person there is a risk of rejection. For example, when a person has a kidney transplant they must take drugs to stop their immune system from attacking it.

In therapeutic cloning, a human embryo would be cloned from the patient. Stem cells could then be taken from this embryo and used to treat the patient's disease. The cloned embryo would have the same genes as the patient, so cells from it would not be rejected. This is therapeutic cloning – creating a human embryo only to make stem cells. It is not cloning to produce a new human adult.

Who objects to stem cell cloning?
Anti-abortion groups are against stem cell research. They say that it means human embryos are destroyed. 'The destruction of human embryonic life is unnecessary for medical progress, as alternative methods of obtaining human stem cells and of repairing and regenerating human tissue exist and continue to be developed.' (The Coalition of Americans for Research Ethics).

People also object to stem cell research because of their religious beliefs. Some people believe that it is the same as killing a child, because human life begins at the moment of fertilization.

---

a Explain the difference between reproductive and therapeutic cloning.

therapeutic cloning would be cloned from the patient where as Reproductic- cloning the cells come from another person and there a change of rejection but they take drugs to stop stop the immune system from attacking it.

b Underline an argument from the article that cloning of human embryos should not be done because it is unnatural or wrong.

c Put a ring around an argument that cloning of human embryos to produce stem cells should be done because it is the best outcome for the majority of people involved.

d Use different colours to shade in one example of each of the following in the article:

fact   opinion   theory   speculation

1 Infections are caused by harmful microorganisms invading your body.

a Use these words to complete the headings in the table.

bacterium	fungus	virus

Type of microorganism	*Virus*	*Bacterium*	*fungus*
Size	less than 1 μm	1–5 μm	over 50 μm
Examples of diseases caused	flu, colds, AIDS, measles	tonsillitis, cystitis, TB	athlete's foot, thrush, ringworm

b Your body has natural barriers to reduce the risk of harmful microorganisms getting inside.

Use these words to complete the sentences.

microorganisms	reproduce	acid	destroy	chemicals

➔ Your skin produces _*chemicals*_ that make

it difficult for microorganisms to _*reproduce*_.

➔ Tears contain chemicals that _*destroy*_

microorganisms.

➔ Your stomach makes _*acid*_ ,

which destroys most _*Microorganisms*_ .

c In suitable conditions microorganisms can reproduce rapidly.

Highlight or draw a (ring) round the conditions that would let microorganisms grow quickly.

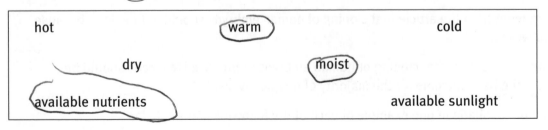

hot   (warm)   cold

dry   (moist)

(available nutrients)   available sunlight

d In the right conditions a bacterium can divide into two every twenty minutes.

Complete the table to show how many bacteria would grow from one bacterium in three hours.

Time		Number of bacteria	Generation
Hours	Minutes		
0	0	1	1
0	20	2	2
0	40	4	3
1	00	8	4
1	20	16	5
1	40	32	6
2	00	64	7
2	20	128	8
2	40	256	9
3	00	512	10

e Use these words to complete the diagram showing how microorganisms can make you ill.

cell	damage	poisons	symptoms	toxins

Some bacteria produce

Poisons called

toxins .

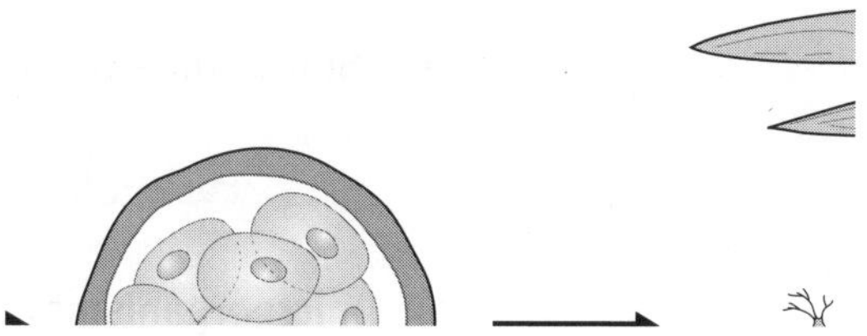

3 days

Harmful microorganisms

cause damage

to body cells. This causes

disease Symptoms

and makes you feel ill.

A Cell infected
with a virus makes many
copies of it.

The viruses burst
open the cell.

2 Your body has ways of fighting infections.

a Your immune system defends you against invading microorganisms.

White blood cells destroy microorganisms by engulfing and digesting them, or by producing antibodies.

Choose words from this list to complete the sentences.

toxins	antibodies	digest	red	white	engulf	infect	markers

➜ Cells have their own particular _Markers_ on the outside.

➜ Foreign markers are recognized by _White_ blood cells in your blood.

➜ Some white blood cells make chemicals called _antibodies_ that stick to markers on microorganisms.

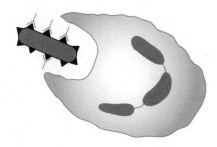

➜ Other white blood cells _engulf_ and _digest_ marked cells.

20

3 Each white blood cell makes only one kind of antibody. A different antibody is needed for each different type of microorganism.

a Only the correctly shaped antibody can recognize a particular type of microorganism.

On the diagram below:

➜ label one white blood cell, one bacterium, and one antibody

➜ draw the correct antibodies on the three bacteria shown at **X**

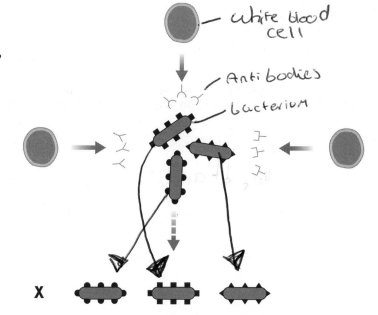

– white blood cell

– Antibodies

– bacterium

X

b It may take a few days for your body to fight off a new infection. Explain what is happening in your immune system during this time.

Killing some Back of the ba· Making the Antibodies to destroy

The infections

c Once your body has made antibodies to recognize a particular organism, it can make them again very quickly when needed.

Use these words to complete the sentences.

antibodies	immune	markers	memory	microorganisms

➜ After an infection, special white blood cells called __Memory__ cells stay in your blood.

➜ If the same type of microorganism gets into your body again, these cells recognize the
Markers
~~Microorganism~~ immediately.

➜ These white blood cells multiply quickly and produce the right ~~Markers~~ _antibodies_ .

➜ The antibodies destroy the __Microorganism__ before they make you ill.

➜ Your __Immune__ system protects you from getting the same infection again.

21

4 Vaccinations can provide protection from some diseases caused by microorganisms.

a Describe how vaccination works by explaining what happens at each of these stages.

➔ A small amount of a dead or inactive form of a disease-forming microorganism is injected into a child.

Small amount of disese mos are put into your body + dead or inactive forms are used so you don't get the disease itself, sometimes just parts of the mos are used

➔ A few years later the child is infected with living microorganisms of the same type.

White blood cells recognise the foreign mos. they make the right antibodies to stick to the mos

➔ The child does not get ill.

The Mos are destroyed before they can Make you ill.

b The virus that causes flu (influenza) is always changing. A different vaccine is made each year to fight the current form of the virus.

Draw the shape of the antibody needed for virus B. Explain why you could catch flu from virus B even if you were immune to flu from virus A.

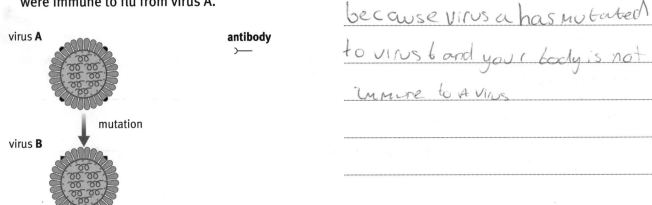

virus **A**

antibody

mutation

virus **B**

because virus a has mutated to virus b and your body is not immune to A virus

c AIDS is a disease caused by a virus called HIV. It is difficult to develop a vaccine against HIV, because the virus damages the immune system and has a high mutation rate.

➔ People infected with HIV become ill with other infections more often than people who are not infected with HIV. Explain why.

because theyre immune system is weaker than others

➔ Why does the high mutation rate of HIV make it difficult to develop a vaccine against it?

because the Hiv would just change its gene by the time the vaccine is ready.

5 Vaccination can never be completely safe. Individuals have varying degrees of side-effects from a vaccine. Risks need to be weighed against potential benefits.

a Look at the table below. Highlight or underline in one colour this sentence and the **benefits and costs to the person having the vaccination**.

b Using a second colour, highlight or underline this sentence and the **benefits and costs to society as a whole**.

Benefits	Risks/costs
Less chance of getting the diseases.	Chance of temporary vaccination side-effects.
Less chance of long term disease complications.	Chance of long-term vaccination side-effects.
Fewer cases of disease means NHS savings on treatments.	Costs of vaccinations.
Less risk to the sick, the old and the very young from an epidemic of the disease.	Other costs, such as loss of work by people and sick pay.

c Vaccination policy has to consider both the issues involved and the views held by different people.

Read these statements and answer the questions.

> 1 It is every child's right to be protected from infectious disease.
>
> 2 No child should be denied vaccination without serious thought about the consequences for the child.
>
> 3 No child should be denied vaccination without serious thought about the consequences for the whole community.
>
> 4 Consent from the parent or guardian should always be obtained before giving a vaccine.

» What is the main issue being considered by these statements?

Should ~~children~~ all children been given vaccines compulsary

» Which statement do you disagree with? Explain why.

4) because if a lot of people decline then then they a more changes that people without get it.

d To prevent epidemics of infectious diseases, a high percentage of the population must be vaccinated.

> Statistics for April 2003 to March 2004 showed 80% of two-year-olds had been given the MMR jab, down from 82% in 2002 to 2003. The figures show that in southeast London, just 62% have been immunized.
>
> The World Health Organisation (WHO) says 95% of toddlers should have the jab to protect them against the diseases. The figures are also well below a peak of 92% coverage seen in 1995–96, before controversial research claimed the MMR jab was linked to autism.

➔ Measles is very contagious; it spreads fast. Explain why the WHO wants 95% of two-year-olds to have the MMR jab.

because even if it spread it will die out because only 5% can get affected

➔ Explain why, if this trend continues, south-east London is at risk of a measles outbreak.

because th

e The WHO target of 95% could be met if the government made vaccination of young children compulsory.

Summarize the arguments for and against compulsory vaccinations in the table.

Arguments for compulsory vaccination	Arguments against compulsory vaccination
Less chance for getting the disease fewer	Children could get small side effect children could get Long-time side effect Cost of vaccination Other of cost, such as loss of work by people and sick pay

f Measles is a more widespread and serious disease in Africa. Explain how this might affect parents' decisions about childhood vaccinations.

6 Chemicals known as antibiotics can kill bacteria and fungi, but not viruses.

a Show with a tick ✓ in the last column which of these illnesses could be treated successfully with antibiotics.

Illness	Type of microorganism	Antibiotic treatment
septic wound	bacterium	✓
athlete's foot	fungus	✓
common cold	virus	
cystitis	bacterium	✓
tuberculosis	bacterium	✓
influenza	virus	

b Some bacteria in a population of bacteria will be more affected by an antibiotic than others.

Show what this means by completing the diagram and notes.

1 Disease-causing bacteria have multiplied in the body. The patient feels ill.

2 Antibiotic treatment immediately _remove_ some of the bacteria. The patient improves.

3 Longer antibiotic treatment kills _the remaining_ the bacteria. The patient is cured.

7 Random changes (mutations) in the genes of microorganisms sometimes lead to varieties that are less affected by the antibiotic.

a Explain, in terms of microorganisms, what is meant by natural variation.

they When certain individuals are better suited to their environment

b Complete the diagrams and notes to explain how antibiotic-resistant bacteria can grow.

1 An infected patient is treated with antibiotics.

2 The patient begins to feel better and forgets to take any more antibiotics.

3 The ones that are few survive change their dna mutate to survive the antibiotics

c A doctor's surgery has these notices in it. Explain the reason for each notice.

Make sure you always finish your course of antibiotics.	
Please don't ask for antibiotics for a cold.	
We only prescribe antibiotics when you really need them.	

8   New drugs are first tested for safety and effectiveness.

a Explain the difference between these tests.

⇒ Safety tests find out _if there is any side effect_

⇒ Effectiveness tests find out _is it does any improvent to the body_

b Drugs are tested in several stages. Put ticks ✔ to show what is being tested at each stage.

Stage	Tests on . . .	How safe the drug is	How well the drug works
1	human cells grown in the lab	✓	✓
2	animals	✓	
3	healthy human volunteers	✓	
4	people with the illness		✓

c People have different views about how drugs should be tested.

Explain why, if a drug works on cells grown in the laboratory, it is then tested on animals.

because it might have side effect in other thing than cells.

e New medical treatments are tested using human trials.

Use these words to complete the table.

blind	double-blind	open

Description	Type of trial
Both doctor and patient know who is taking the drug.	*open*
Only the doctor knows who is taking the drug.	*blind*
Neither patient nor doctor knows who is taking the drug.	*double-blind*

f Draw lines to match each of these terms used in clinical trials with the correct description.

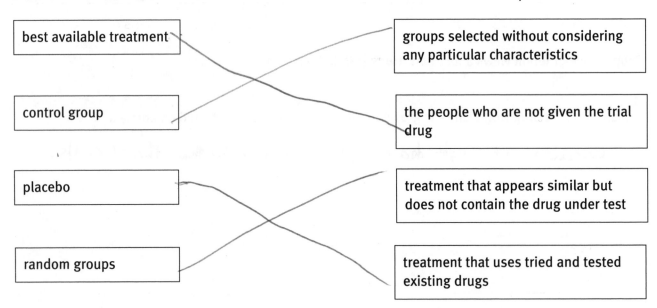

best available treatment	groups selected without considering any particular characteristics
control group	the people who are not given the trial drug
placebo	treatment that appears similar but does not contain the drug under test
random groups	treatment that uses tried and tested existing drugs

g In most human trials, the effects of the new drug are compared with the effects of the best treatment available at the time. In a few cases the new drug is compared against a placebo.

When do doctors agree that it is right to use a placebo in a drug trial?

_____

_____

_____

h Explain why a random double-blind trial is considered the best type of clinical trial.

_____

_____

9   Heart muscle cells, like other body cells, need a good supply of oxygen and food.

a Blood is circulated round your body through arteries, capillaries, and veins.

Label these diagrams.

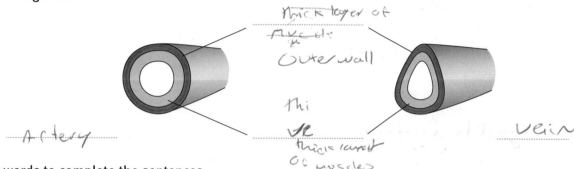

thick layer of muscle outer wall

thi

VE
thick layer of muscles

Artery

Vein

b Use these words to complete the sentences.

arteries	veins	food	valves	oxygen	thick

The heart pumps blood carrying ___oxygen___ and ___food___ to all parts of the body.

___Arteries___ carry blood away from the heart.

They have ___thick___ walls to carry the pulse of blood pushed through them at every heartbeat.

Blood is carried back to the heart in ___veins___.

They have ___valves___ to stop blood flowing backward.

c Fat can build up in the coronary arteries.

Draw a diagram to show this in the first box. Then add notes to the flow chart to explain how this might lead to a heart attack.

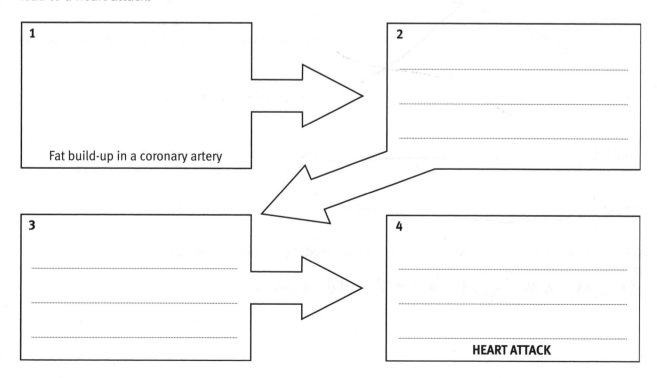

1

Fat build-up in a coronary artery

2

3

4

HEART ATTACK

10 Heart disease is more common in the UK than in non-industrialized countries.

a

> Heart disease is not normally caused by microorganisms. It is usually caused by lifestyle factors and/or genetic factors.

Which piece of information from the box tells you that:

➔ heart disease runs in some families

     Genetic factors

➔ you can reduce your risk of heart disease

     life style factors

➔ heart disease is not an infectious disease

     ca not caused by microorganism.

b Lifestyle factors that may cause heart disease include poor diet (particularly a lot of saturated fat), stress, cigarette smoking, and high levels of alcohol consumption.

Draw lines to match each lifestyle factor with the main reason that it is bad for your heart.

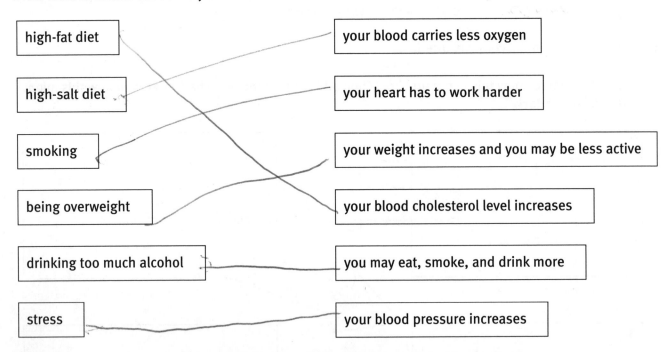

high-fat diet	your blood carries less oxygen
high-salt diet	your heart has to work harder
smoking	your weight increases and you may be less active
being overweight	your blood cholesterol level increases
drinking too much alcohol	you may eat, smoke, and drink more
stress	your blood pressure increases

c Write down one factor not shown above that can help **reduce** your risk of heart disease.

The eat more good cholestrol like nuts.

11 The lifestyle factors that increase the risk of heart disease can be identified through large-scale (epidemiological) studies.

a Draw lines to match each word with the correct meaning.

| cause | | steps that explain how a factor causes an outcome |

| mechanism | | there is evidence that a factor produces an outcome |

| correlation | | there is evidence of some link between a factor and an outcome |

b Large-scale studies show that both sales of suntan cream and cases of sunburn increase in the summer.

Highlight or underline the correct interpretation(s) of this piece of evidence.

➤ There is a correlation between sales of suntan cream and cases of sunburn.
➤ Suntan cream causes sunburn.
➤ There is a correlation between cases of sunburn and summer.
➤ Summer causes cases of sunburn.

c Graphs are a good way of showing how different factors affect each other.

Complete the sentences to describe what the graph shows.

As the body-mass index _____Incruse_____ the

number of heart attacks _____Incuse_____. There is

a _____Lrnk_____ between these two factors. But

this does not prove that a high body-mass index is

the _____A cavse_____ of a heart attack

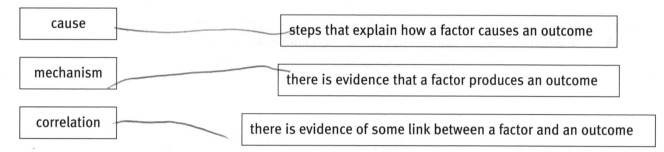

y-axis: number of heart attacks
x-axis: body-mass index

d Read this BBC news report and complete the sentences below.

Carrying fat around the waist can quadruple the risk of diabetes and heart disease, say University of Birmingham scientists.

'Fat cells which develop round the waist pump out chemicals. These chemicals are needed in small amounts. But in large doses they can cause damage, raising blood pressure and increasing cholesterol in the bloodstream.'

➤ There is a correlation between _____diebetes and High blood pressure and increasing cholostrol_____

➤ Scientists think that this is caused by _____Life style cheices_____

12 Good epidemiological studies can provide convincing evidence for new scientific claims.

a A good epidemiological study looks at a lot of people (a large sample). Two studies looked at men's health and the amount and type of exercise they took.

Study A: in South Wales between 1984 and 1988, of 3000 men aged 49–65
Study B: in Newcastle between 1985 and 1990, of 1000 men aged 30–60

⇨ Explain which study should be the most reliable.

⇨ Explain why it is good to have two separate studies.

b If a study is comparing two groups of people, the groups should be of similar sorts of people (well matched).

A study is looking at whether owning a cat reduces heart disease. List some characteristics that should be matched in a group of cat owners and a group of people who do not own a cat for the study.

c Explain the importance of each step in the 'peer review' process in checking that new scientific claims are reliable.

⇨ A new claim is published with results to support it.

⇨ Other scientists working in the same field check the scientific work.

⇨ More evidence is gathered and results are replicated by other scientists.

⇨ The replicated results support the claims, which are then accepted by other scientists.

Life on Earth – Higher

1 All the different species on Earth evolved from simple living things. This includes species that are now extinct.

a Draw a line to match each of these key words with their meanings.

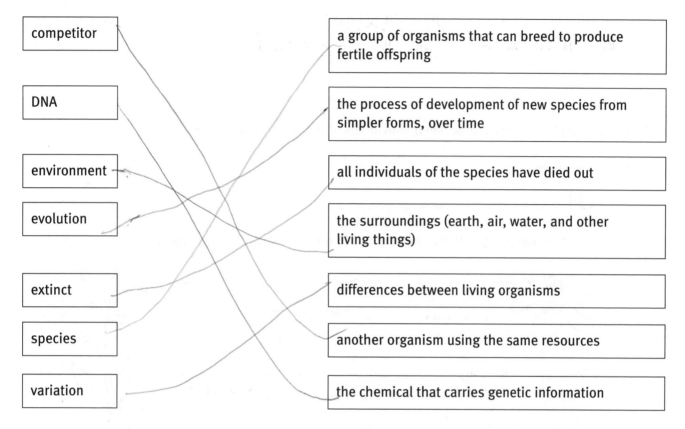

competitor	a group of organisms that can breed to produce fertile offspring
DNA	the process of development of new species from simpler forms, over time
environment	all individuals of the species have died out
evolution	the surroundings (earth, air, water, and other living things)
extinct	differences between living organisms
species	another organism using the same resources
variation	the chemical that carries genetic information

b Use these words to complete the sentences.

DNA   evidence   evolutionary   extinct   fossils   similarities   simple

Living organisms have many _Similarities_. Scientists have studied dead bodies found in

Fossils. This evidence shows that living organisms have changed over time, and that many

organisms have become _extinct_.

Scientists have an explanation for this fossil _evidence_. They say that all life on Earth has evolved

from very _simple_ organisms. More recent evidence from studying _DNa_ agrees

with this explanation. This new evidence is being used to work out where different species fit into the

evolutionary tree.

c What are the two main sources of evidence for the theory of evolution?

⮕ _Fossils_   ⮕ _DNA_

d Give two examples of similarities common to all living things.

They all breath oxygen, they all eat food.

e Give two examples of variation between individuals of the same species.

Eye colour

Weight

f Look at the diagram of two plants that have grown in different environmental conditions.

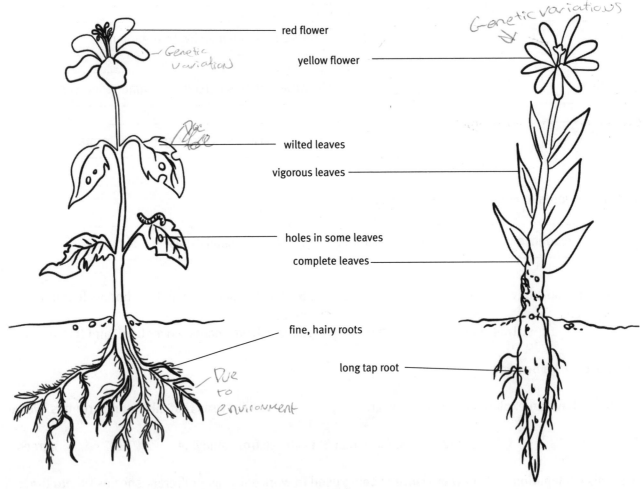

red flower

Genetic variation

yellow flower

Genetic variations

wilted leaves

vigorous leaves

holes in some leaves

complete leaves

fine, hairy roots

long tap root

Due to environment

⮕ Highlight or <u>underline</u> in one colour this sentence and **the genetic variations shown in the diagram.**

⮕ Highlight or <u>underline</u> in another colour this sentence and **variations that are mainly due to the environment.**

⮕ Which sort of variations can be passed on to the offspring?

Petal colour

2 Evolution happens by the process of natural selection.

a Number the sentences to explain the process of natural selection.

4 In later generations, there are more individuals with the successful features.

3 Some successful features are passed on to the offspring.

1 Some individuals have features that help them survive when competition gets tough.

2 These successful individuals have a better chance of surviving and reproducing.

5 Individuals of a species show variation – they are not all identical.

b Most wild primulas are yellow, with occasional pink ones. A plant breeder developed a bright red primula.

Explain the process of selective breeding by completing the sentences. Use the words below to help your explanation.

bred them together   variation   breed   red flowers   pink primulas   deep pink flowers

➡ The wild primulas showed _variation That most primulas were yellow or occassionally pink_

➡ The plant breeder selected _Red flower deep Pink flowers Pinker primulas and NPE P pink Primola_

➡ They then took these plants and _bred them together bred them together_

➡ The breeder selected offspring with _red flower Red Flower_

➡ Over several generations the breeder _was made a ste primulas with Red Flowers._

c Give an example of a living organism that has recently changed due to natural selection.

Human beings

d What is the evidence for the change?

Our ancester did not surviv the other apes did not survive also The Fossils prove it

e Explain how the change happened.

Because flower that dest

3 Life on Earth began about 3500 million years ago. It developed from molecules that could copy themselves.

a Fill in the missing word.

Most living things on Earth today contain _RNa_ . This is a molecule that can copy itself.

b Use these words to complete the sentences.

3500 million	conditions	copy	Solar System	theory

The first living things were molecules that could _copy_ themselves. They first appeared on

Earth about _3500_ _million_ years ago. Some scientists think that these molecules

came from somewhere else in the _Solar System_ . Another _theory_ is

that life started on Earth. At that time the _conditions_ might have been just right to produce

these molecules.

c Suggest a reason why there can be two different theories of where life on Earth came from.

because theory was no evidence they no evidence for the theorys

d Multicellular organisms evolved millions of years later. They became specialized in many different ways. Gradually the huge variety of life evolved over more millions of years. Over that time some organisms became extinct.

Suggest a reason for these extinctions.

because they could not get that what they need to get it it needs to survive

e The conditions on Earth over millions of years have affected the variety of organisms here today. Explain this, using the words **conditions, characteristics, survived** in your answer.

f Complete the sentence.

If at any stage the conditions on Earth had been different, evolution by natural selection would have

produced _a multicellular organism that would survive the environment_ .

4 In multicellular organisms, their cells need to communicate.

a Multicellular organisms have evolved nervous systems. These are made up of nerve cells (**neurons**) linking **receptor cells** to **effector cells**.

Label the diagram and give another example from the human nervous system.

for example **1:** light receptors in the eye   neurons in the optic nerve   muscles around the pupil

        2: ....

....

b Sally put 25 maggots on a wire mesh in a Petri dish. She covered half the dish in black card. She put wet cotton wool under half the wire mesh. The diagram shows the dish.

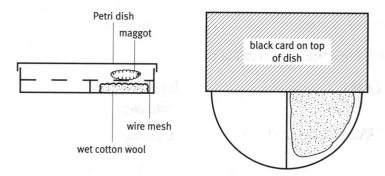

After 10 minutes Sally counted the maggots in each quarter of the dish.

	Dark and dry	Dark and wet	Light and dry	Light and wet
Number of maggots	3	19	2	1

i What two stimuli (environmental conditions) are the maggots responding to?

Dark and wet

ii Explain how the maggots' response would help them survive in their natural environment.

because it would help them look for food and protect from other creatures

c In humans and other vertebrates the nervous system is coordinated by a central nervous system.

Underline or highlight the two structures that form the central nervous system.

bones   <u>brain</u>   eyes   glands   muscles   <u>sensory cells</u>   spinal cord

d Multicellular organisms have also evolved hormonal systems. These communicate through chemicals.

Put arrows onto the flow chart and give another example from the human hormonal system.

A gland produces a hormone, which is released into the blood.		The hormone causes a response in another part of the body.

The blood carries the hormone around the body.

for example **1:** The pancreas produces insulin.   Insulin circulates round the body.   Insulin affects cells to control blood sugar.

2: _____   _____   _____

e Complete the table to compare the nervous and hormonal communication systems in humans.

	Nervous system	Hormonal system
Tissues involved	neurons, spinal cord, brain	glands
How signals are carried	cells	blood
Response fast or slow?	fast	slows
Response lasts long or short time?	short time	long

f The nervous and hormonal communication systems are also involved in maintaining a constant internal environment. Keeping conditions constant is called homeostasis.

Use these words to complete the sentences below.

constant   feedback   nerve cells   nervous system   sweating

Body temperature is kept constant by a _feedback_ mechanism. A control centre in the brain

senses any rise in body temperature. Impulses travel along _nerve cells_ to switch on various

mechanisms that cool the body, such as increasing _sweating_ . If the body temperature falls, the

control centre sends messages to switch on mechanisms that warm the body. The _nervous_

system is involved in keeping the body temperature _constant_ .

g Explain why the secretion of insulin by the pancreas is an example of a hormonal system that is involved in homeostasis.

because the insulin is carried by blood and the effect last for a long time.

5 Humans evolved from our hominid ancestors. As more evidence is found, scientists look again at exactly how this evolution happened.

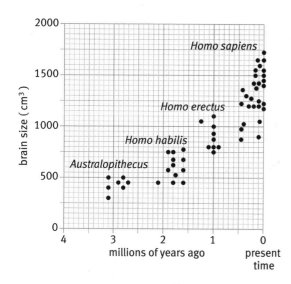

a The graph shows the brain size of some different hominids. Explain why the scientist used data from several members of each species in this graph.

So they could find the average size of the Brain of the species

b Use the graph to suggest a theory about brain size and survival.

The bigger the Brain the more change of survival.

c All the hominid species except *Homo sapiens* (humans) became extinct. Suggest an explanation for their extinction.

The other hominid could not handle the competition for food

d Suggest an explanation of why *Homo sapiens* is the only surviving hominid species.

because we have a bigger brain we as know how to get food.

e Use the terms in the box to label the diagram showing how humans evolved separately from early hominids.

common ancestor	early hominids	modern humans

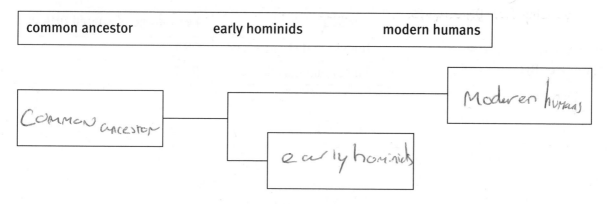

6 Explanations of evolution of life on Earth need to account for all the current data.

a Draw a line to match each of these key words with the correct definition.

data	a measurement, or experimental record
evidence	a way of accounting for some facts
explanation	facts collected together
observation	facts tending to support or disprove an explanation

b Look at the diagrams of feet.

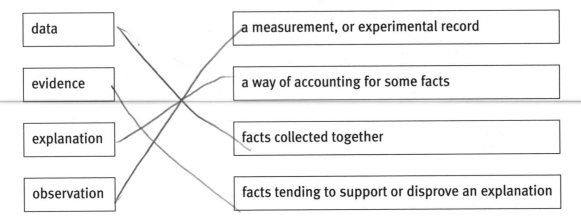

chimpanzee   gorilla   human

☐ **A**

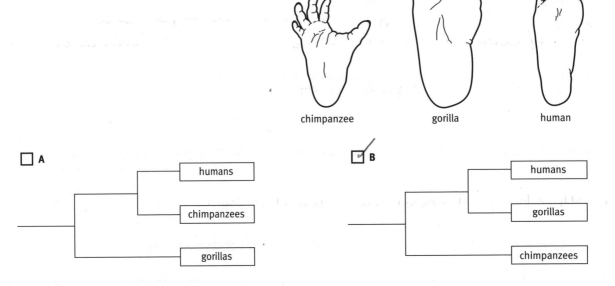

humans

chimpanzees

gorillas

☑ **B**

humans

gorillas

chimpanzees

i Tick ✓ the diagram (A or B) that best explains the data.

ii Describe the evidence that supports this explanation.

The feet size is similar and the toe is nearer to the others toes

c The table shows data from DNA studies.

Does this support your chosen explanation (A or B)?

Species	Percentage of shared DNA (%)
human and chimpanzee	98.4
human and gorilla	97.7

B No

d Read the text in the box and answer the questions.

> Darwin made many notes about the living organisms around him. He observed that different species often had some characteristics in common. He observed that members of a single species showed considerable variation. He observed that members of one species are in competition for food and space. He thought about these observations and realized that if some members of a species were better suited to survive they would have more offspring. He then explained how evolution of life on Earth could have happened through characteristics that were inherited and a process of natural selection.
>
> Mendel collected data from plant breeding experiments. From this evidence he was able to describe how characteristics were inherited.

i Highlight or underline in one colour this sentence and **Darwin's explanation of evolution**.

ii Highlight or underline in another colour this sentence and **an observation that is accounted for by this explanation**.

iii In what way did Darwin show imagination and creativity?

He thought about these observation;

iv In what way did Mendel's work support Darwin's explanation of evolution?

because plant also do Natural Selection.

v When Darwin published his work, not all scientists thought he was right. Suggest some reasons why there was disagreement.

because in those lots of people were religious and They believed The bible which says Earth was build in six days.

7 The combined effect of mutations, environmental changes, and natural selection can produce new species.

a Read the text in the box and answer the questions.

One famous example of natural selection is the peppered moth. These moths rest on tree trunks. Birds feed on the moths. Before the Industrial Revolution, in the eighteenth century, most tree trunks were covered in pale lichens. The moths were pale in colour.

In industrial areas, pollution from burning coal killed many lichens. The tree trunks became darker. In 1849 a dark form of the peppered moth was discovered in industrial Manchester. This dark form of the moth quickly became the most common in polluted areas.

i Suggest a reason why dark moths may have survived better in industrial areas.

because they were more camaflaged

ii A dark moth arose from a mutation. Explain what this means.

a change happen in one of the genes

iii How could the mutation have been passed on to its offspring?

By DNA

b Use the example of peppered moths to explain how a new species can evolve. Finish these sentences.

➜ Mutations produce *a change in it is DNA*

➜ A change in the environment means that *change could be a advantage*

➜ Natural selection means that *speices survive and the other dan't*

➜ Over many generations *a new speices is formed*

8   Living organisms depend on the environment and other species for their survival.

a Draw a line to match each of these key words with the correct definition.

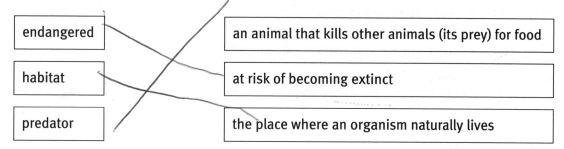

endangered		an animal that kills other animals (its prey) for food
habitat		at risk of becoming extinct
predator		the place where an organism naturally lives

b Look at the food web for the Arctic.

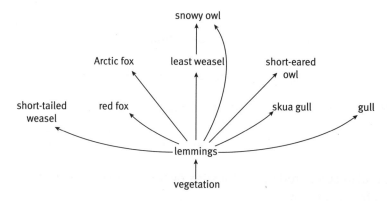

i What group of organisms do all the animals in that habitat depend on?

Producers

ii Explain how the Arctic fox and the snowy owl are in competition.

because they both eat LeMMings

iii Explain what might happen if environmental conditions changed the vegetation.

The Snowy owl The Lemming will die and every species dies

iv Explain what might happen if disease wiped out all the birds.

Nothing appart from They more Lemming

v Explain what might happen if a new predator, such as a wolf, was introduced.

The Lemming might be exint or The Other eomptition could be exint

c A rapid change in the environment may cause a species to become extinct.

Suggest two other changes in a habitat that might cause one species to become extinct.

➜ _Rapid climate change_

➜ _diesease._

9   Scientists measure and record the species found in a particular environment. This helps them understand the ways in which different organisms depend on each other.

Scientists often measure and record species by sampling.

Read the account about studying hedgerows and answer the questions.

We surveyed 50 randomly selected hedge sites. At each site we measured a 10 m section of hedge.
We recorded:
- the average maximum height
- the average maximum width
- evidence of hedge cutting
- percentage cover of different plants
- the number of fruits present in autumn
- percentage of areas showing gaps
- the use of the land next to the hedge

a Explain what is meant by 'random' selection of sites.

b The survey compared sites with each other. It also compared the same sites over time. Explain why it was important to use a standard procedure for all the recording.

c It took a long time to survey a 10 m stretch of hedge. Explain why such a large sample of each hedge was used.

d One scientist suggested that the variety of different plant species in a hedge tells the probable age of the hedge. Suggest how you could collect evidence to support this hypothesis.

10 Biodiversity is the variety of life on Earth. Reduced biodiversity can reveal environmental problems.

Read the information in the box and answer the questions.

Hedgerow management, dormice, and biodiversity

A study of farm hedgerows found that:

➜ Most farm hedges are now cut by machine every year. It used to be done by hand about every 6 years.

➜ This machine cutting leads to hedges with fewer plant species.

➜ This machine cutting leads to hedges with less fruit, nuts, and seeds.

➜ There has been a 70% decrease over 25 years in hedgerow sites with dormice.

➜ Wider and taller hedges contained more dormice.

➜ The number of dormice was a good indicator of the biodiversity in the hedge.

---

a What effect has machine cutting had on the plants and animals living in hedges?

It has cut some are has made the species go down in number

b Suggest some explanations why machine cutting has led to fewer dormice.

because machines cut cut everything,
but kept by hands we Don't cut everything

c Suggest why the number of dormice is a good indicator of the biodiversity in a hedge.

d Suggest a practical way of managing farm hedges that would encourage biodiversity.

11 Extinctions can be caused by human activity, either directly or indirectly.

a Draw a line to match each of these key words with the correct definition.

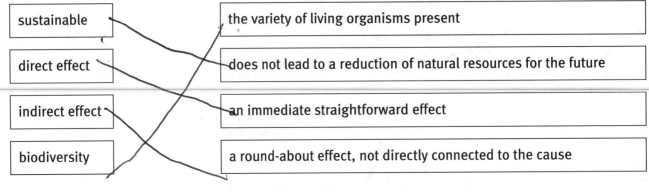

sustainable	the variety of living organisms present
direct effect	does not lead to a reduction of natural resources for the future
indirect effect	an immediate straightforward effect
biodiversity	a round-about effect, not directly connected to the cause

b This diagram shows threats to a loggerhead turtle.

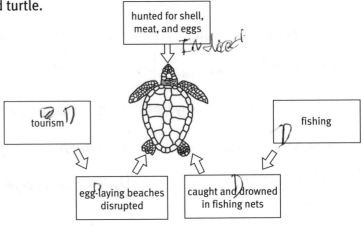

hunted for shell, meat, and eggs *Indirect*

tourism

fishing

egg-laying beaches disrupted

caught and drowned in fishing nets

On the diagram:

➔ highlight or <u>underline</u> in one colour this sentence and **threats caused by direct human activity**

➔ highlight or <u>underline</u> in another colour this sentence and **threats caused by indirect human activity**

c The loggerhead turtle is an endangered species. Explain what this means.

Species that are nearly extinct

d Fill in the table below with examples from the box. Then add your own examples.

> ➔ Changes to farming practice destroyed the meadow grass that the large blue butterfly needed for breeding, so it became extinct in the UK.
> ➔ The great auk was hunted to extinction for food and feathers.
>
> ➔ Sailors introduced rats, cats, and dogs to the island of Mauritius. A century later the flightless dodo was extinct.
> ➔ The passenger pigeon was tasty to eat and was hunted to extinction.

Extinctions caused by direct human activity	Extinctions caused by indirect human activity
great auk was hunted to extinction for food and feathers	*Changes to farming practise destroyed the meadow grass that was the large blue butterfly*
	the dodo

12 We need to maintain biodiversity and use the environment in a sustainable way.

We are using up fossil fuels, and global warming is on the increase. Future generations may need different species of plants, animals, and microorganisms to provide food, fuel, clothes, and medicines.

Add notes to the diagram to explain why we need to maintain biodiversity, and use the environment in a sustainable way now, for the benefit of future generations.

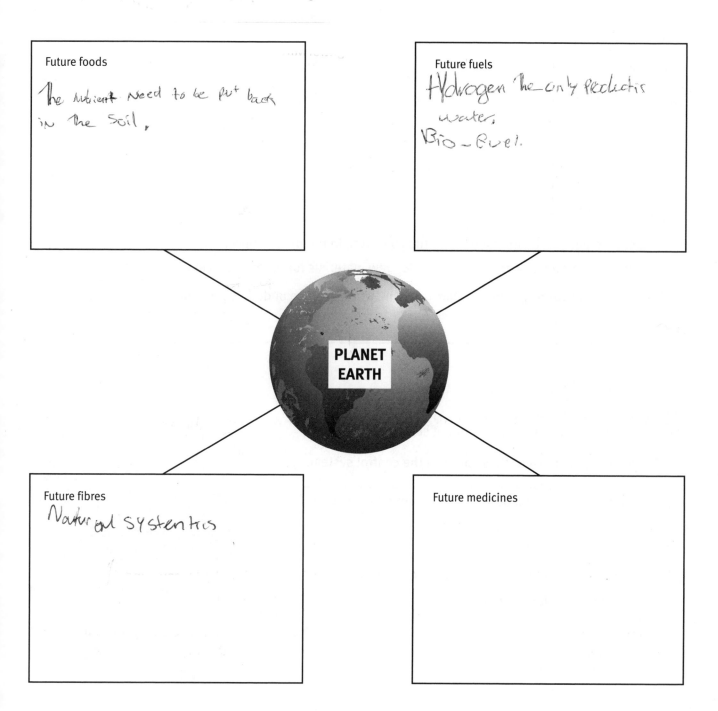

Future foods

The nutient Need to be put back in the Soil,

Future fuels

Hydrogen the only predictis water.
Bio-fuel.

Future fibres

Natural Systentics

Future medicines

Homeostasis - Higher

1 In and out of cells

Molecules move in and out of cells all the time.
This happens so that conditions inside the cell are kept steady.

a Complete the diagram of a cell to
show chemicals moving in and out.
Use words from the list.

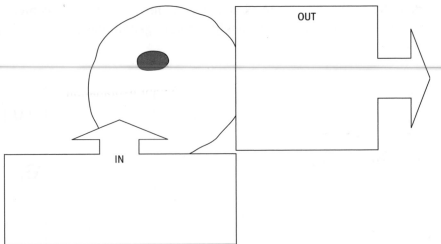

oxygen

carbon dioxide

urea

digested food
(e.g. glucose)

water

b Use coloured pencils to <u>underline</u> the words you wrote in the boxes above.

➜ toxic waste products = **red**

➜ raw materials for respiration = **blue**

➜ raw materials for other chemical reactions in cells = **green**

Some words may be underlined in more than one colour.

2 Control systems

This incubator has an automatic control system.
It keeps conditions inside the incubator at steady levels.

a Label the diagram to show these parts of the control system:

stimulus	receptor	processing centre
effector	response	

b Your body also has control systems for homeostasis.

Complete the definition of homeostasis:

Homeostasis means _____

_____

c Body temperature is controlled by homeostasis.

Give one example of another factor that must be kept steady
in the body.

_____

_____

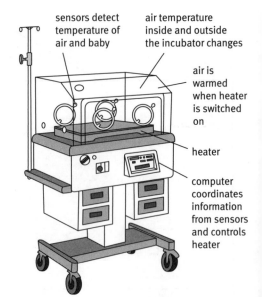

sensors detect
temperature of
air and baby

air temperature
inside and outside
the incubator changes

air is
warmed
when heater
is switched
on

heater

computer
coordinates
information
from sensors
and controls
heater

3   Negative feedback

a Write a definition for negative feedback.

....

....

b Explain why negative feedback is important in control systems.
Use the example of an incubator to help your explanation.

....

....

4   Antagonistic effectors

An incubator has only one effector. It has a heater to warm up the air.

Your body temperature control system has effectors to both:

➡ warm your body up *and*

➡ cool your body down

a Explain why body temperature control is an example of antagonistic effectors.

....

....

b Explain why having antagonistic effectors is an advantage in a body control system.

....

....

5   Balancing body temperature

Animals gain and lose heat from their environment.

Complete this sentence.

To keep a steady body temperature, input and output must be balanced so that:

heat .... = heat ....

6   Body temperature control systems

This text describes how your body controls its temperature.

Read the text then answer the questions.

Receptors in your skin detect changes in the temperature of the air around you. Receptors in your brain detect changes in the temperature of your blood. This information is passed to the temperature control centre in the brain – called the hypothalamus.

The hypothalamus coordinates all the information from the receptors. It automatically triggers effectors in the body to respond to changes in body temperature. These effectors are sweat glands and muscles.

If body temperature rises too high:
- sweat glands increase production of sweat
- muscles in blood vessels supplying the skin relax

These responses work to lower body temperature.

If body temperature becomes too low:
- sweat production is reduced
- muscles in blood vessels supplying the skin contract
- skeletal muscles contract rapidly, causing shivering

These responses work to raise body temperature.

---

a Shade each of the boxes next to these terms a different colour.

stimulus ☐   receptor ☐   processing centre ☐   effector ☐   responses ☐

b Use these colours to highlight or <u>underline</u> parts of the temperature control system in the text above.

c Explain *how* these responses help to raise or lower body temperature:

- producing more sweat ....

....

- shivering ....

....

d Blood vessels carrying blood to the skin have muscles in the walls.
These muscles can be contracted (vasoconstriction) or relaxed (vasodilation).

The diagrams show vasoconstriction and vasodilation.
Explain *how* these responses help to control body temperature.

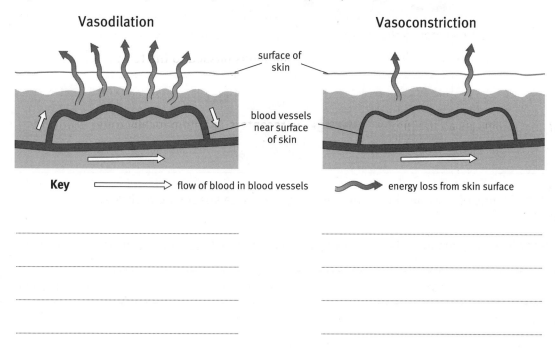

---

7 Enzymes

Enzymes are a very important group of chemicals found in living things.

a What type of chemical molecule are enzymes made of? ....

Hydrogen peroxide is a waste product of chemical reactions in many living cells. It is poisonous, so it is broken down by cells into water and oxygen. The enzyme catalase is very important in breaking down hydrogen peroxide.

A student measured how fast a sample of hydrogen peroxide was broken down with and without catalase.

The diagram shows their equipment. The graph shows their results.

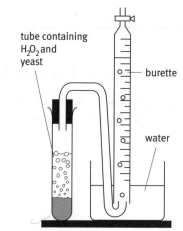

b Where is the catalase in this experiment?

....

c What effect does catalase have on the breakdown of hydrogen peroxide?

....

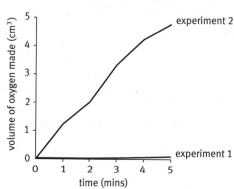

d Complete the boxes to explain how the enzyme catalase works.

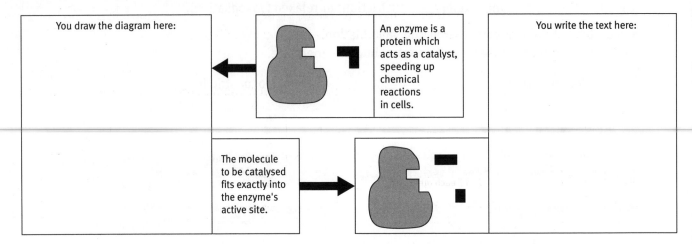

| You draw the diagram here: | | An enzyme is a protein which acts as a catalyst, speeding up chemical reactions in cells. | You write the text here: |

| | The molecule to be catalysed fits exactly into the enzyme's active site. | |

e There are tens of thousands of enzymes in the human body. Each one speeds up a different chemical reaction.

Explain why an enzyme can speed up only one particular reaction. Use these key words in your answer:

| active site | shape | substrate | lock and key |

....

....

....

8   Enzymes and temperature

Enzyme reactions are affected by temperature. Complete the notes to explain why this happens. Use these key words in your notes:

| collisions   optimum   denatured |

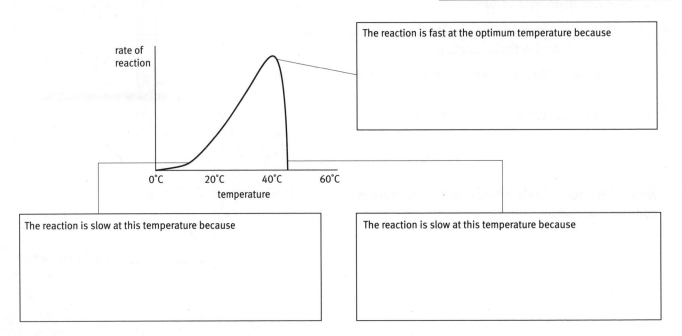

The reaction is fast at the optimum temperature because

rate of reaction

0°C   20°C   40°C   60°C
temperature

The reaction is slow at this temperature because

The reaction is slow at this temperature because

9 Diffusion

Molecules are always moving. In gases and liquids they move randomly.
The molecules bump into each other and spread out. This happens by diffusion.

a Tom was watching television when his mother sprayed some air freshener at the other side of the room.
After a few minutes Tom could smell the scent of the air freshener. The diagrams explain what happened.

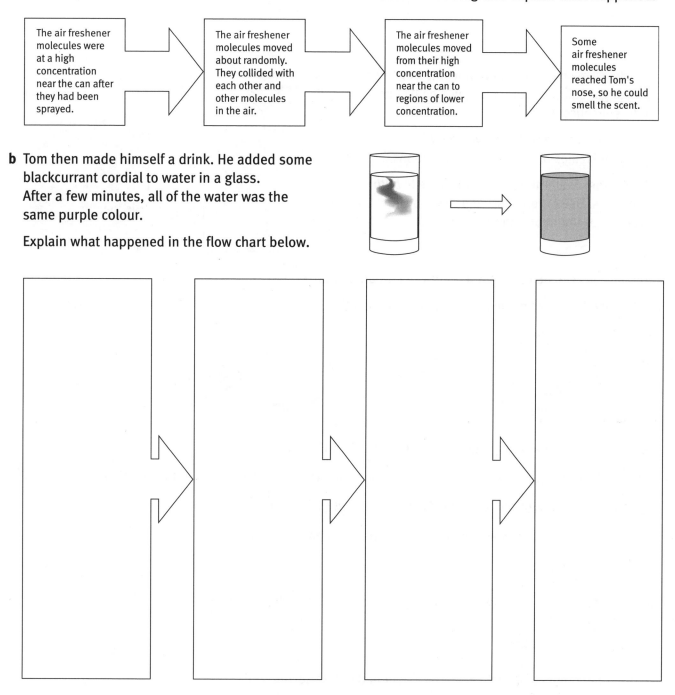

The air freshener molecules were at a high concentration near the can after they had been sprayed.

→

The air freshener molecules moved about randomly. They collided with each other and other molecules in the air.

→

The air freshener molecules moved from their high concentration near the can to regions of lower concentration.

→

Some air freshener molecules reached Tom's nose, so he could smell the scent.

b Tom then made himself a drink. He added some blackcurrant cordial to water in a glass.
After a few minutes, all of the water was the same purple colour.

Explain what happened in the flow chart below.

c Write down three examples of chemicals which move in or out of cells by diffusion.

10 Osmosis

Osmosis is a special case of diffusion.

a Write down a definition of osmosis.

....

b The diagram shows a semi-permeable bag filled with sugar solution. It is sitting in a beaker of a different concentration of sugar solution.

Draw arrows to show the movement of water molecules between the two solutions.

(Remember, water molecules will move in *both* directions. Use different sized arrows to show the *overall* movement of water molecules.)

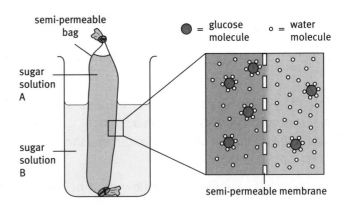

c Animal cells can be damaged if their water balance is upset.

A red blood cell was put in a very dilute sugar solution.

⮂ The solution was more dilute than the contents of the cell.

⮂ There was a higher concentration of dissolved molecules in the cell than in the solution.

⮂ Water entered the cell by osmosis, so the cell swelled up and burst.

A second red blood cell was put into a concentrated sugar solution.

Write three points to explain what has happened to this second cell.

⮂ ....

⮂ ....

⮂ ....

11 Active transport

a Complete the sentences to describe active transport. Use words from the list.

active transport	high	low	energy	passive

Diffusion does not need any _____ from the cell. It is a _____ process. Diffusion

only moves molecules from a _____ to a _____ concentration. Some molecules move into

cells by a process called _____   _____ .

b Colour the numbered parts in the diagram to match the words in these sentences.

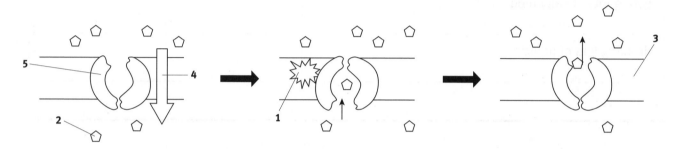

Active transport uses energy₁ to move molecules₂ across cell membranes₃ from low to high
concentrations. This is against their concentration gradient₄. The energy is used to change the
shape of a carrier protein₅ in the membrane.

---

12 Water balance

Water moves in and out of your body every day. These inputs and outputs must be balanced for your cells to
work properly.

Complete the table to show how your body gains and loses water.

List ways your body gains water	List ways your body loses water
1.	1.
2.	2.
3.	3.

13 Kidneys balance the body's water levels

The kidneys make urine. The amount of water in urine changes.

Complete the table to show how different conditions affect the concentration of your urine.

Conditions	Concentration of blood	Level of water in urine	Concentration of urine
cold day, staying inside	low	high	dilute
hot day, playing sport outside			
eating lots of salty food			
drinking lots of liquids			

14 Kidneys get rid of the body's waste

a Use these words to label the diagram on the left.

urine	tubules	kidney

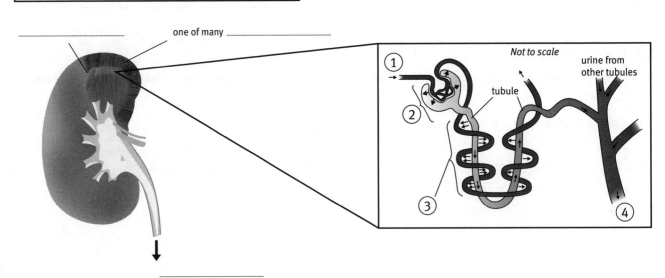

one of many

b Put the correct number in each box to describe what is happening in the diagram on the right.

☐ Small molecules are **filtered out** of the blood into the tubule, including urea, glucose and water. Salt ions are also filtered out.

☐ **Reabsorption** of useful ions of salt, glucose and water molecules.

☐ Blood flows to the tubule.

☐ Waste molecules form urine which is stored in the bladder.

15 Water balance is controlled by ADH

The amount of water in urine is controlled by a hormone called ADH.

a Complete the control system diagram to explain how this works.

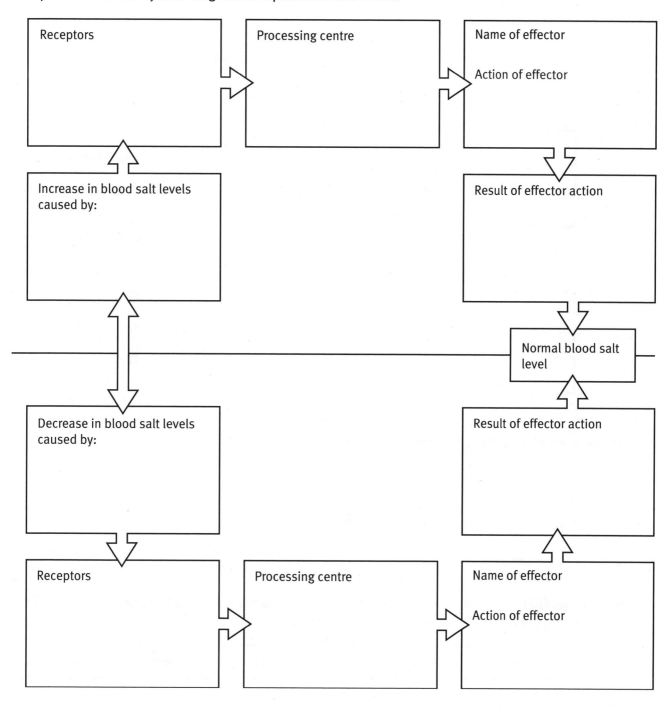

b Explain why water balance by ADH is an example of negative feedback.

....

....

....

16 Extreme environments

a Complete the list of four factors which must be kept steady in the body:

temperature

_____   _____

_____   _____

b Complete the graph to show the normal range of core body temperature.

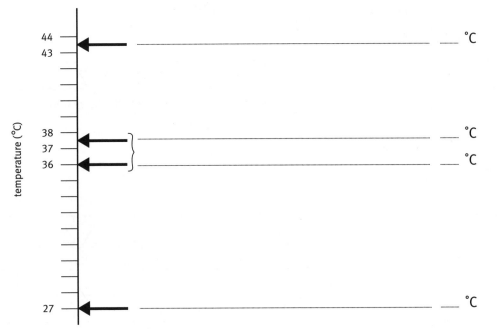

44 ← .... °C	
43	lower limit of survival
	upper limit of survival
38 ← .... °C	range of normal body core temperature
37 }	
36 ← .... °C	36
	47
	27
	37.5
27 ← .... °C	

temperature (°C)

c Write down the parts of the body that make up the core.

_____

d Write down two parts of the body that are extremities.

_____ and _____

e Very harsh conditions can stop homeostasis from working properly. Complete the sentences to explain how very hot temperatures can affect the body. Use words from the list.

core	decrease	up	rise	low	increases	dehydrated

Condition: Very hot weather.

Homeostasis demands: Body temperature goes _____. Sweating _____.

This may cause body water levels to fall too _____ if the lost water is not replaced.

The person may become _____. If this happens sweating may _____.

This will cause body temperature to _____ even higher. If the _____ body temperature

becomes too high, the body's temperature control systems stop working.

f Explain how very cold weather can affect the body's homeostasis systems.

Condition: Very cold weather.

Homeostasis demands: ....

....

....

....

g Explain how two different sports can put demands on the body's homeostasis systems.

Sport: ....

Homeostasis demands: ....

....

....

....

Sport: ....

Homeostasis demands: ....

....

....

....

17 Heat stroke

a Complete the definition of heatstroke.

Heatstroke is ....

....

b Complete the information leaflet about heatstroke.

> **Avoiding heatstroke**
>
> People may suffer heatstroke as a result of
>
> ➔ ....
>
> ➔ ....
>
> Look out for the following symptoms:
>
> ➔ ....
>
> ➔ ....
>
> ➔ ....
>
> You should take the following actions to cool the patient:
>
> ➔ ....
>
> ➔ ....
>
> ➔ ....
>
> ➔ ....

18 Hypothermia

a Complete the definition of hypothermia.

Hypothermia is when ....

Body heat cannot be ....

b Complete the information leaflet about hypothermia.

Avoiding hypothermia

People on mountains may become chilled as a result of

➔ ....

➔ ....

Look out for the following symptoms:

➔ ....

➔ ....

➔ ....

➔ ....

You should take the following actions to warm the patient:

➔ ....

➔ ....

➔ ....

➔ ....

Do not give a hot water bottle because

➔ ....

Do not give food or alcohol because

➔ ....

---

19 Drugs and homeostasis

Certain drugs can upset the body's water level control system.

They do this by affecting the production of ADH.

a Complete the sentences to explain the role of ADH. Use words from the list.

| increases | more | decreases | reabsorbed | pituitary | kidney | hormone |

Water is filtered out of the blood in the .... Some of this water is .... to

keep the body's water level balanced.

ADH is a .... It is made by the .... gland in the brain. ADH causes the

kidneys to reabsorb .... water.

When the body's water levels are too low, ADH production ....

If the body's water levels are too high, ADH production ....

b Complete the table to show the effects that certain drugs have on the body's water balance.

Drug	Effect on ADH production	Urine production	Effect on body water balance
caffeine	decreased	greater volume of dilute urine made	dehydration
alcohol			
Ecstacy			

c Ecstacy also increases sweat production. Explain how this may cause serious disruption to the body's temperature control system.

....

....

....

Growth and development – Higher

1 New cells in plants specialize into cells of roots, leaves, or flowers.

a Draw a line to match each of these key words with their meanings.

development	the sequence of growth and reproduction
growth	increase in size
life cycle	increase in complexity
specialized	adapted for a purpose

b In plants and animals, cells become specialized to do a particular job.

Use these words to complete the boxes and add another example.

organs   tissues

1

cells: can specialize to do a particular job

2

.... :
groups of specialized cells

3

.... :
groups of tissues with a particular function

For example, in plants: xylem   flower

Another example from plants is: ....

c Give an example of a specialized plant cell and explain how it is suited to do its particular job.

....

....

2 Cells in an early human embryo can develop into any sort of cell, but soon cells become specialized to form a particular type of tissue.

a The diagram shows a human egg being fertilized and starting to develop.
Choose words from this list to complete the labels.

| embryo | sperm cell | nucleus | zygote | egg cell | cytoplasm | fertilization |

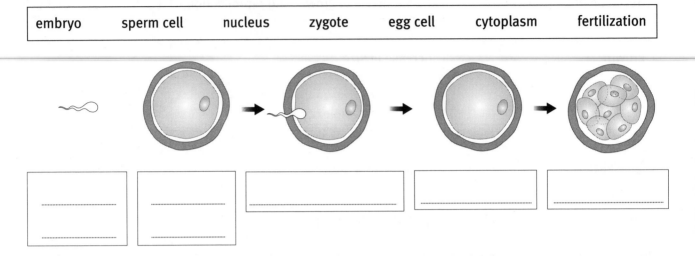

b Complete this sentence.

A human zygote must contain .... for making all the different types of cells in the

human body.

c Up to the 8-cell stage, the cells in the developing embryo are not specialized.
Complete the notes below the diagram to describe an important characteristic of these embryonic stem cells.

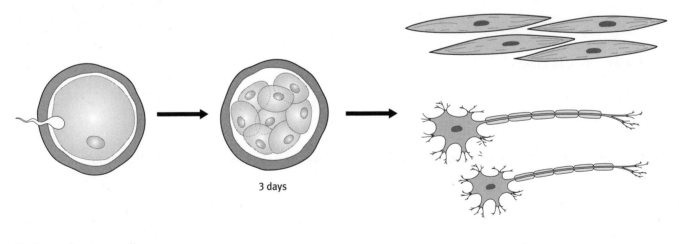

3 days

Embryonic stem cells can ....

....

3 Unlike most animal cells, some plant cells remain unspecialized and can develop into any type of plant cell.

a Flowering plants can continue growing all their lives. Use these words to complete the sentences.

| longer | meristem | roots | shoots | stems | taller | thicker |

Plants have unspecialized cells called _____ cells. These cells mean that plants

can go on growing at the tips of the _____ and _____, and in the width

of the _____. These unspecialized cells can produce different tissues so that plants can

continue to grow _____, with _____ stems and _____ roots,

throughout their life.

b Meristem cells mean that a piece cut off from a plant can, in the right conditions, grow into a complete new plant. Draw lines to match up these half sentences.

Plants that are genetically identical . . .	. . . clones of a plant with desirable features.
Producing plants from cuttings produces plants . . .	. . . are known as clones.
Cuttings can be used to produce . . .	. . . that are identical to the parent.
Producing plants from seeds produces plants . . .	. . . that vary, with some characteristics from each parent.

c Fill in the table to compare growth in humans and plants.

Characteristic	Humans	Plants
period of growth in size		
ability to repair slight tissue damage		
ability to regrow lost limb or branch		
ability to grow whole organism from a cutting		

4 During growth, cells divide by mitosis producing two new cells identical to each other and to the parent cell.

a Use these words to label these diagrams of two typical cells.

chromosomes	cytoplasm	nucleus	organelles

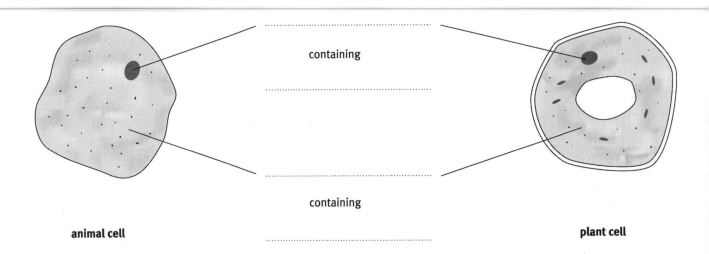

containing

....

containing

....

animal cell   **plant cell**

b Living organisms grow by making new cells. Complete the diagram to show the stages in the cell cycle in a cell with four chromosomes (in 2 pairs).

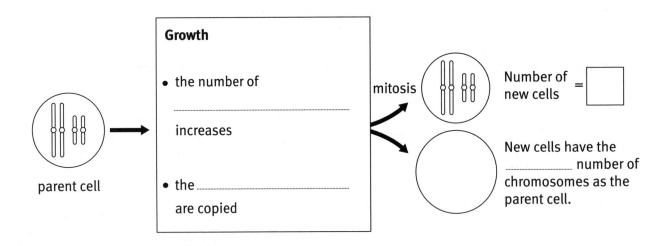

c Complete this sentence.

Cell division for growth is called .... There are .... new cells produced

which are genetically .... the parent cell.

5 When forming gametes (sex cells), cells divide by meiosis producing four new cells with half the number of chromosomes of the parent cell.

a Cell division that produces gametes is called meiosis. Complete the diagram showing meiosis in a cell with four chromosomes (in two pairs).

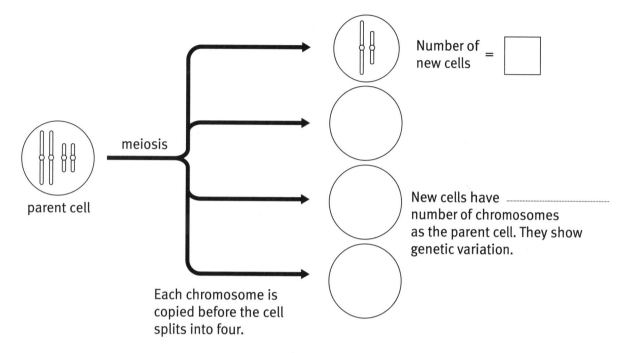

Number of new cells = ☐

New cells have _____ number of chromosomes as the parent cell. They show genetic variation.

parent cell

meiosis

Each chromosome is copied before the cell splits into four.

b Complete this sentence.

Cell division to form gametes is called _____ . There are _____ new cells

produced, with half the number of chromosomes of the parent cell. They show genetic _____ .

c In sexual reproduction, a male gamete fuses with a female gamete. The gametes each have half the number of chromosomes of the parent organism. The zygote has a set of chromosomes from each parent.

Fill in the number of chromosomes in each human gamete.

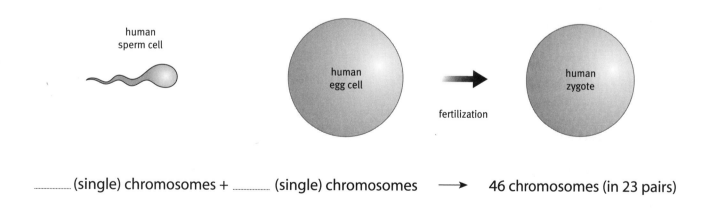

_____ (single) chromosomes + _____ (single) chromosomes ⟶ 46 chromosomes (in 23 pairs)

d Fill in the table (with ✓ and ✗) to compare mitosis and meiosis.

Feature	Mitosis	Meiosis
cell division for growth		
cell division for gamete production		
four new cells formed		
two new cells formed		
new cells identical to parent cell		
new cells have half the chromosomes of the parent cell		
new cells show genetic variation		
new cells genetically identical		

e Complete each of the sentences with one of these words.

meiosis	mitosis

⮕ The zygote divides by _____ to form an embryo.

⮕ Cells in the ovary divide by _____ to form egg cells.

⮕ Cells in the root tip divide by _____ as the root grows.

⮕ Plant cells divide by _____ to form pollen grains.

6 Genes control growth and development within the cell.

a The chromosomes in the cell nucleus are made up of many genes. Each gene is a length of DNA. DNA molecules have a double helix structure.

Use these words to complete the diagram.

double helix	chromosomes	genes	nucleus

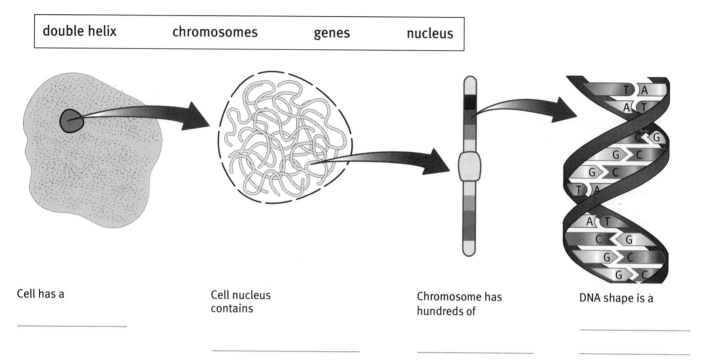

Cell has a

....

Cell nucleus contains

....

Chromosome has hundreds of

....

DNA shape is a

....

b A DNA molecule is made up of two chains connected by pairs of bases.

Draw a diagram to show the two-dimensional structure of DNA (like a ladder).

➔ Label a pair of bases (a rung).
➔ Label the two chains (the ladder sides).

7 **Chromosomes are copied when the two strands of each DNA molecule separate and new strands form alongside them.**

a Look at the diagram below showing how a double strand of DNA is copied to form two identical strands. Complete these sentences.

- There are .... different bases in DNA.
- Base A always pairs with base ....
- Base C always pairs with base ....

b Label the bases in the two new strands of DNA shown.

c Complete the sentences under each diagram to show how two exact copies of the DNA are made.

1

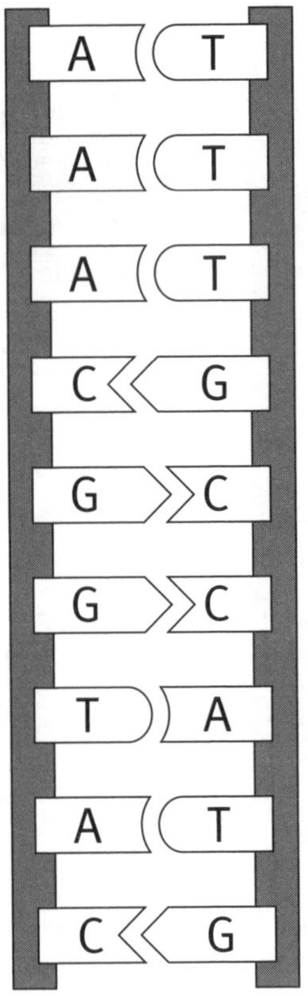

2

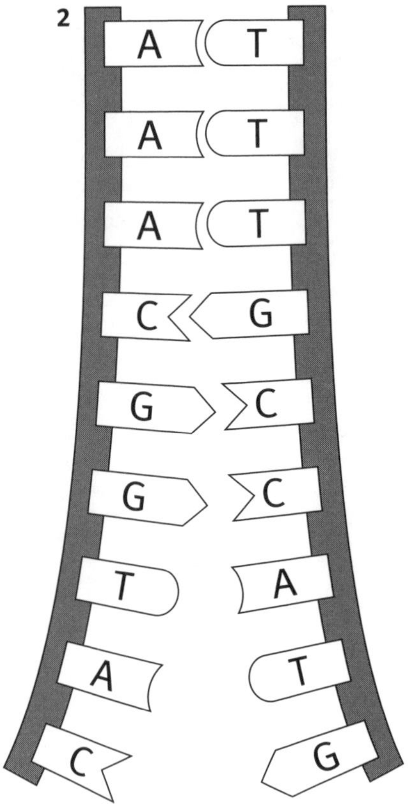

3

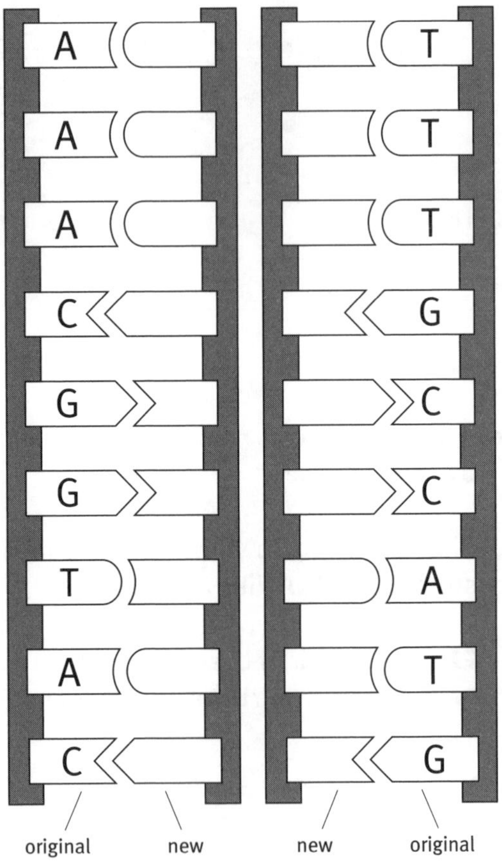

.... along the DNA are joined in matching pairs.

Weak bonds between the bases split. The DNA opens into .... strands. Free bases in the cell pair with the bases on each open strand.

The result is two DNA molecules. Each molecule is half new. The .... pairs are in the same order as in the original DNA.

8   Body cells in an organism all contain the same genes. But many genes in a particular cell are not active because the cell produces only the proteins that it needs.

a Each gene carries the instructions for a different protein. Proteins have many different functions. Draw a line to match each protein with its description.

amylase		a structural protein of hair and nails, hard
chlorophyll		a structural protein of ligaments, strong
collagen		a structural protein of skin, stretchy
elastin		a digestive enzyme
insulin		a green pigment that absorbs light energy
keratin		a hormone involved in controlling blood sugar

b Draw a line to match each protein with the cells that make them.

amylase		hair cell
chlorophyll		salivary gland cell
keratin		green leaf cell

c Some proteins are found in every type of cell. Complete this sentence.

All cells need energy from respiration. The genes that code for enzymes used in respiration are

switched on in _____ cell.

71

d Cell specialization means that a cell produces only the proteins that it needs.

Complete the table with ✔ and ✗ to show which genes are active in these human cells.

Cell	Gene coding for keratin	Gene coding for amylase	Gene coding for a cell membrane protein
salivary gland cell			
hair cell			
heart muscle cell			

e Complete these sentences.

➡ Embryonic stem cells are unspecialized – all the genes are switched _____ .

➡ As the cells specialize, some genes are switched _____ .

➡ A cell will only make proteins from genes that are switched _____ .

➡ The _____ that a cell makes control how it develops.

9 Stem cells have the potential to produce cells needed to replace damaged tissue. Genes can sometimes be reactivated in cloned cells to form cells of different tissue types.

a Complete the diagram to explain how embryonic stem cells could be used to make different types of tissue for medical treatment.

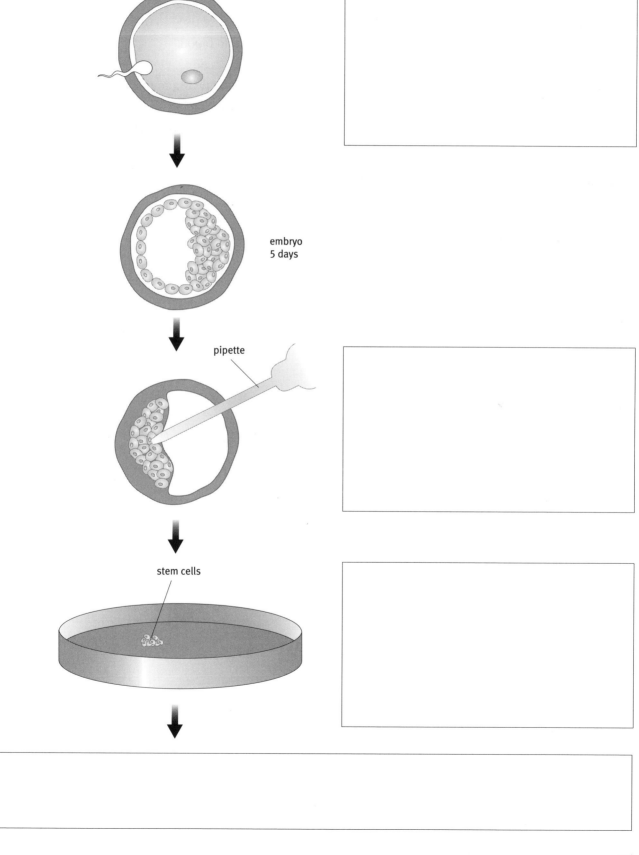

b Read the newspaper report and answer the questions.

What does 'cloning' mean to you? Is it cloned people in scary movies? Well, don't panic. Scientists are cloning, but not to make new people. They want to find cures for illnesses.

Some diseases are caused because cells in the body stop working properly. Scientists hope to clone cells from healthy people. These healthy cells would then be given to the patient.

But there is a problem. Growing cells from adult humans is not very easy. Adults have only a few types of stem cells. These are unspecialized cells that can keep dividing. In adults they are used to replace worn out cells like skin and blood.

Scientists are trying to find a way of 'despecializing' adult stem cells. Then adult stem cells could be used to make any type of cell.

➔ Explain how the first step in therapeutic cloning differs from the production of embryonic stem cells shown on page 13.

....

....

....

➔ Explain why adult stem cells are only of limited use.

....

....

....

➔ Explain why therapeutic cloning using an egg cell would not be necessary if scientists could find a way to 'despecialize' adult stem cells.

....

....

....

10 During protein production a copy of the gene is made in the nucleus. This copy carries the DNA instructions to the cytoplasm, where the protein is assembled.

a The diagram shows the steps necessary for an active gene to make a protein. Carefully label the diagram using these words.

| cytoplasm | amino acids | DNA | mRNA | nucleus | ribosome | protein |

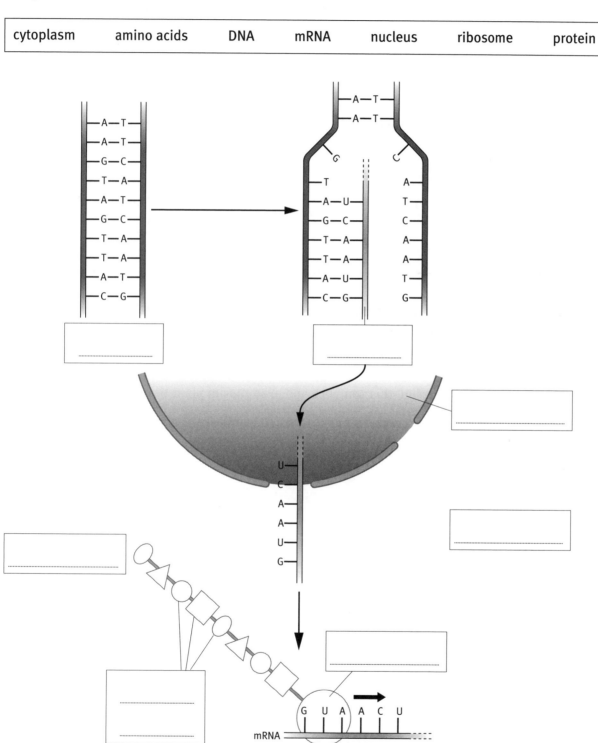

each triplet codes for one amino acid

b Proteins are made in organelles in the cytoplasm called ribosomes. During protein production a copy of the gene is made in the nucleus (mRNA). This carries the DNA instructions to the cytoplasm, where the protein is assembled.

Put numbers in the boxes to show the sequence.

☐ | In the cytoplasm, a ribosome joins amino acids together in the order coded for by the mRNA.

☐ | In the nucleus, the DNA double strand separates.

☐ | A molecule of mRNA forms along the DNA strand, with bases pairing with the DNA bases.

☐ | The order of amino acids decides what protein is made and its particular characteristics.

☐ | The small mRNA molecule passes through the pores of the nuclear membrane.

---

11 If the hormonal conditions in their environment are changed, unspecialized plant cells can develop into a range of other tissues or organs.

a As a plant grows, cells specialize into tissues which arrange themselves into organs.

➔ Highlight or underline in one colour this sentence and **examples of plant tissues** in the list below.

➔ Highlight or underline in another colour this sentence and **examples of plant organs** in the list.

➔ Circle the **unspecialized plant cell** named in the list

| flowers | leaves | meristem | phloem | roots | xylem |

b Use these words to complete the sentences.

| tissue | roots | hormones | meristem | clone |

Unspecialized plant cells can make any kind of _____ the plant needs. Rooting powder can be

used to encourage cut shoots to form _____. Rooting powder contains plant hormones.

The _____ cause the new cells produced by the _____ cells in the shoot to develop

into roots. The cutting then grows into a complete plant which is a _____ of the parent.

c Explain the advantages of taking cuttings as a way of reproducing particularly good plants.

_____

_____

12 Phototropism increases a plant's chance of survival.

a Complete the sentence.

The rate at which a plant grows depends on the rate of photosynthesis. The rate of photosynthesis

depends on the availability of carbon dioxide, water, and ——————————————————.

b Highlight or <u>underline</u> the correct description of phototropism.

➔ the process of using energy from light to make glucose
➔ the bending of plant shoots towards light
➔ the bending of plant shoots away from light
➔ providing extra light to get maximum plant growth

c Rooting powder contains plant hormones called auxins. Auxins increase the rate of plant growth. They are also involved in phototropism.

Complete the diagrams of the plants to show the way that the plant shoots would be growing.

Add notes to explain how auxins are involved.

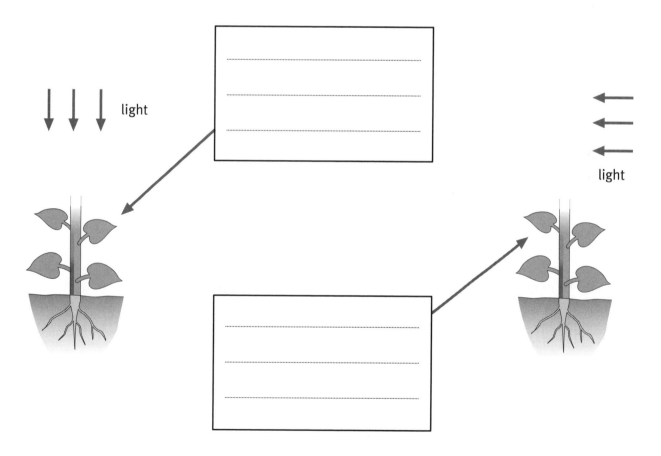

d Explain why phototropism is likely to increase a plant's chance of survival.

——

——

1 Animals respond to stimuli in order to keep themselves in favourable conditions.

a A stimulus is a change in the environment of an organism. Look at the examples of behaviour below.

→ Highlight or <u>underline</u> in one colour this sentence and **the stimuli**.

→ Highlight or <u>underline</u> in another colour this sentence and **the responses**.

> Woodlice prefer dark places; they move away from light.
>
> Bacteria living in the gut move towards the highest concentration of food.
>
> An earthworm rapidly withdraws into its burrow if pecked.
>
> A resting housefly flies takes off as soon as it sees any fast movement nearby.
>
> If an octopus sees a predator, it releases a cloud of 'ink' and moves away quickly.

b Simple reflexes are automatic. They give responses that help an animal survive and reproduce. Complete this list.

Simple reflex behaviour helps an animal to:

→ find ....

→ escape from ....

→ find a ....

→ avoid .... environments

c Simple animals rely on reflex actions for most of their behaviour. This means they cannot adapt their behaviour, or learn from experience. Explain the disadvantages of this.

....

....

....

d Fill in the words that best fit these descriptions. You will find all the words you need on this page.

→ an action or behaviour caused by a stimulus ....

→ a change in the environment that causes a response ....

→ the way an organism responds to a stimulus or situation ....

→ a response to a stimulus that is involuntary ....

2 Simple reflexes produce rapid involuntary responses.

a Simple reflexes are responses that you do not think about or learn. Look at the list below. Highlight or <u>underline</u> the examples of simple reflexes.

➔ Your pupils get smaller in bright light.

➔ You answer a question.

➔ Your eyes water on a windy day.

➔ A newborn baby grasps at anything put in her hand.

➔ A goalkeeper saves a goal.

➔ You breathe faster when you run.

b Humans and other mammals have very complex behaviour, but simple reflexes are also important for their survival. Complete the descriptions of human reflexes in the tables, and add some more examples.

Adult reflex	Stimulus	Response
gagging	touch to back of the throat	closing of throat
pupil		

Newborn reflex	Stimulus	Response
grasping	touch to palm of hand	
stepping		walking movement of legs
sucking		

c Many newborn reflexes are present for only a short time after birth. Explain why they increase a young baby's chances of survival. Use these words in your explanation.

behaviour	experience	learn

....

....

....

3 Some complex organs have receptors and effectors.

a Multicellular organisms respond to stimuli through **receptor** and **effector** cells.
Fill in the gaps using the bold words.

detection of stimulus   →   response to stimulus

_____ cell   →   _____ cell

b Different receptors detect different types of stimuli. List some examples of:

➔ receptors that detect changes outside the body _____

_____

➔ receptors that detect changes inside the body _____

_____

c Some receptors are made up of single cells. Others are grouped together as part of a complex organ.
Complete the sentences.

Single-cell receptors are found in human _____, for example _____ receptors.

An example of a human sense organ is the _____, which detects _____.

d Responses to stimuli are carried out by effector organs. Effector organs are either glands or muscles. Use
these words to complete the sentences.

contract	hormones	move	secreting

➔ The effector cells in glands respond to stimuli by _____ chemicals, for example

_____, enzymes, or sweat.

➔ The effector cells in muscles respond to stimuli by causing the muscle to _____ and

_____ a part of the body.

4 Responses are coordinated by the central nervous system (CNS). Sensory and motor neurons carry the signals.

a In mammals the nervous system is made up of a **central nervous system** (the **brain** and **spinal cord**) connected to the body via the **peripheral nervous system**.

Label or colour the diagram to show the parts of the human nervous system printed in bold.

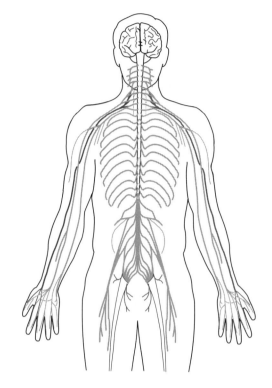

b Neurons are cells in the nervous system that carry nerve impulses. Use these words to label the diagram of a neuron.

| axon | cell membrane | cytoplasm | fatty sheath | nucleus |

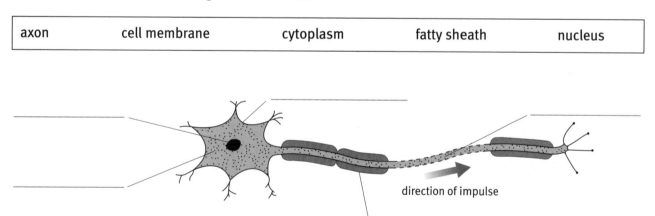

direction of impulse

c Neurons transmit electrical impulses when stimulated. Use these words to complete the sentences.

| axon | electrical | insulates | speed |

When the neuron is stimulated, an _____ impulse travels along the _____ to the

branched ending. Here it connects with another neuron or an effector. Some axons are surrounded by a

fatty sheath, which _____ them from neighbouring cells and increases the _____

of the nerve impulse.

d **Sensory** neurons carry impulses from **receptors** to the CNS. **Motor** neurons carry impulses from the CNS to **effectors**.

Label the diagram using the bold words.

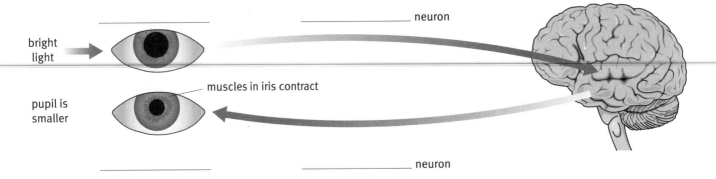

.... neuron

bright
light

muscles in iris contract

pupil is
smaller

.... neuron

e The diagram above shows a reflex arc coordinated by the brain. Reflexes are involuntary actions, and most are coordinated by the spinal cord.

Fill in the flow diagram to describe the route of the nerve impulses when you pick up a hot plate.
(The diagram above will help you.)

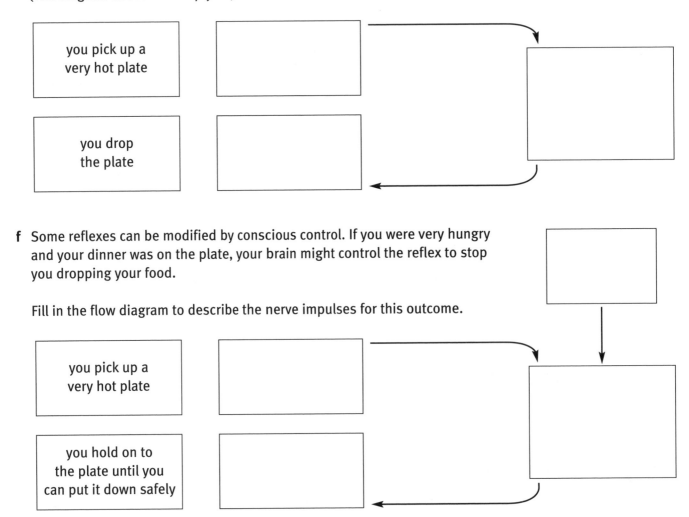

| you pick up a very hot plate | | |

| you drop the plate | | |

f Some reflexes can be modified by conscious control. If you were very hungry and your dinner was on the plate, your brain might control the reflex to stop you dropping your food.

Fill in the flow diagram to describe the nerve impulses for this outcome.

you pick up a
very hot plate

you hold on to
the plate until you
can put it down safely

5 Chemicals released into the synapses transmit nerve impulses from one neuron to the next.

a Synapses are tiny gaps between neurons. Electrical impulses cannot jump across synapses. Chemicals carry impulses between neurons.

Use the diagrams of an impulse crossing a synapse to put the events below in order. Number the boxes 1 to 5.

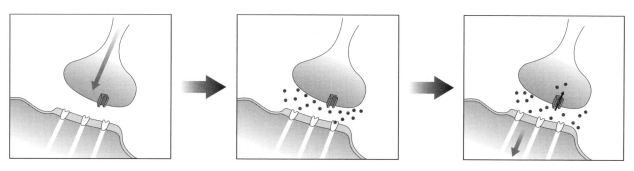

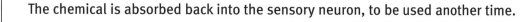

☐ The chemical is absorbed back into the sensory neuron, to be used another time.

☐ An impulse is stimulated in the motor neuron.

☐ A nerve impulse travels along a sensory neuron until it reaches a synapse.

☐ The molecules diffuse across the synapse. They bind to receptor molecules on the membrane of the motor neuron.

☐ The end of the sensory neuron releases a chemical into the synapse.

b The receptor molecules only bind to certain chemicals. Complete the diagrams.

➔ Add receptors of the correct shape to synapse **A**.

➔ Add chemicals carrying the impulse to synapse **B**.

➔ Add arrows to show the direction of the nerve impulse across these synapses.

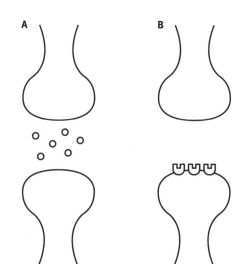

6 Some drugs and toxins affect the way impulses cross synapses.

a Use colours or lines to match up these words with their descriptions.

caffeine	a poison that causes dangerous effects in the body
curare	a medicine or other chemical that causes effects in the body
drug	a stimulant present in tea, coffee, and some soft drinks
Ecstasy	a painkiller used as a medicine
morphine	a chemical that increases nervous activity
painkiller	a chemical that reduces the sensation of pain
stimulant	a poison that blocks transmission of nerve impulses (causing paralysis)
toxin	a drug that has mood-enhancing effects

b Explain how this drug would interfere with the transmission of nerve impulses across the synapse.

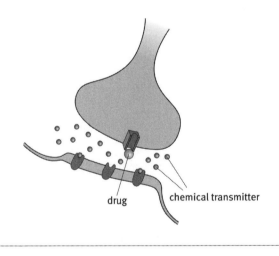

drug   chemical transmitter

c Highlight or colour the drug that would interfere with the transmission of an impulse in the synapse shown in the diagram.

d Explain how this drug would stop the nerve impulse from crossing the synapse.

_____

_____

_____

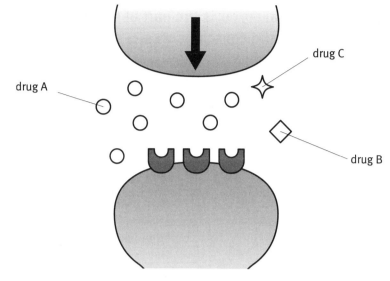

drug A

drug C

drug B

e Some drugs work by blocking the reuptake of the synapse chemical. Serotonin is released at one type of synapse in the brain. It triggers nerve impulses causing feelings of pleasure. The drug Ecstasy (or MDMA) blocks the reuptake of serotonin.

Look at the diagram and answer the questions.

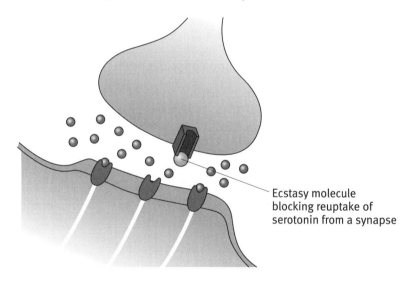

Ecstasy molecule blocking reuptake of serotonin from a synapse

i What effect does the presence of the Ecstasy molecule have on the amount of serotonin in the synapse?

_____

ii What effect does this have on the receptor molecules detecting serotonin?

_____

iii What effect does this have on the activity of this neuron?

_____

f Complete the sentence.
The mood-enhancing effects of Ecstasy are due to the _____ in serotonin concentration at synapses in the brain.

7 The cerebral cortex is the part of the brain most concerned with intelligence, memory, language, and consciousness.

a Choose one of these words to complete the sentence.

hundreds	thousands	millions	billions

The human brain is made of .... of neurons.

b Draw a line to match this word to the best description.

| consciousness |

| being quick to respond, clever |

| caring about others |

| being aware of yourself and your surroundings |

| being wide awake |

c The diagram shows the right side of a human brain that has been cut in half.

⇒ Colour the part most involved in intelligence, memory, language, and consciousness.

⇒ Then complete the description.

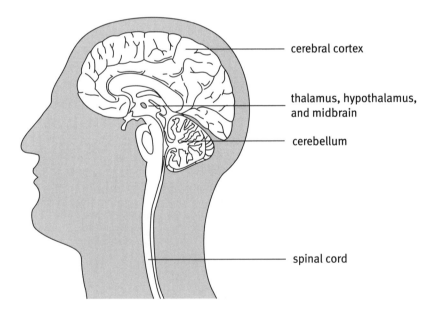

cerebral cortex

thalamus, hypothalamus, and midbrain

cerebellum

spinal cord

The .... is highly folded, giving it a .... surface area.

Different areas are responsible for different .... It is much bigger in ....

than in other mammals.

d Studies of patients whose brains have been partly destroyed by injury or disease tell us about the functions of different areas of the cortex.

Use the information below to decide which area of the brain has been damaged in stroke patients with the symptoms shown in the table.

Symptom	Damaged brain region (A to E)
speech is slurred	
has trouble controlling hand movements	
legs feel numb	

Part of cerebral cortex	Function
A speech centre	talking
B sensory cortex	receiving information from receptors
C motor cortex	voluntary movement
D visual cortex	detecting visual stimuli
E language	understanding language

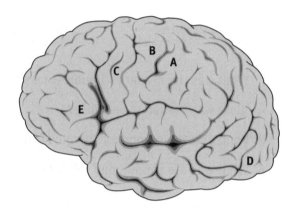

e Regions of the cortex have also been mapped using electrical stimulation during open brain surgery. Explain how this can show the function of different areas of the motor cortex.

....

....

....

f Modern MRI brain scans give information about the function of different parts of the cortex. Explain the advantages of using MRI.

....

....

....

8 Conditioned reflexes can be learned. The final response has no direct connection
to the stimulus.

a Animals can learn to link a new stimulus with a reflex action. This is a conditioned reflex response.
Pavlov showed this in experiments with dogs. Read what he did and answer the questions.

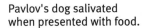

Pavlov's dog salivated
when presented with food.

Pavlov rang a bell while
his dog was eating its food.

After a while the dog salivated
when it heard the bell, even
if no food was around.

i What was the primary stimulus (that originally caused salivation)? _____

ii What was the reflex response? _____

iii What was the secondary stimulus (that Pavlov added)? _____

iv What was the conditioned reflex response? _____

b Complete the sentences.

Salivation is a _____ response connected to food. Salivation and hearing a bell have no

direct connection. In a _____ reflex, the response has no direct connection to the stimulus.

c Conditioned reflexes can increase chances of survival. Read the text in the box and answer the questions.

Many birds feed on caterpillars. Some caterpillars are brightly
coloured and taste nasty (to the birds).
Young birds try eating the caterpillars and learn that they taste
nasty. In future they avoid brightly coloured caterpillars.

i What was the primary stimulus? _____

ii What was the secondary stimulus? _____

iii Which animal had an increased chance of survival? _____

iv Explain how hoverflies increase their chances of survival by having markings that make them look
like wasps.

_____

_____

_____

9   Learning is the result of experience. It creates pathways in the brain that are more likely to transmit impulses than others.

a Draw a line to match each of these key words with their meanings.

adapting	knowledge or skills gained from experience
learning	doing the same thing more than once
neuron pathways	adjusting to new conditions
repetition	strong links between points along connecting neurons

b Human babies learn very quickly. Number these sentences to explain the sequence of events in the brain during learning. One has been done for you.

1	The cortex in a baby's brain is a complicated network of neurons.
	The response is learned.
	Strengthened connections make it easier for more impulses to travel along the pathway.
	A new experience sets up new pathways between the neurons in the cortex.
	Using the pathway strengthens the connections between the neurons.

c Explain why repetition helps you learn a new sporting or musical skill.

....

....

....

10 An animal can adapt well to new situations if it has a variety of potential pathways in the brain.

a Adapting to new situations means learning new skills. Explain why humans are good at learning new skills throughout their life.

b As we get older it becomes harder for the language-processing area in the cortex to make new pathways. Explain how this would affect an adult learning a new language.

c Describe how you could strengthen new pathways in the language-processing area of the cortex when learning a new language.

d Use these words to complete the sentences.

destroyed	development	difficulty	experience

As the brain develops, ... strengthens some neural pathways. Some connections

that haven't been strengthened by experience are .... There is evidence from

studying child .... that children may acquire some skills only at a particular age.

Studies of rare cases of 'feral' or neglected children, who have not learnt any language in early childhood,

show that they have .... learning language skills later in their childhood.

11 Memory means stored and retrieving information.

a Highlight or <u>underline</u> two phrases that together describe memory.

learned behaviour	storage of information	processing of information
retrieval of information	input of sensory information	

b Complete the sentence.

Memory concerning words or language is called _____ memory.

c Memory can be divided into short-term memory and long-term memory.
Complete the table comparing these.

Memory type	How long does it last?	How much can be stored?	An example
short-term			remembering that this row is about short-term memory
long-term			remembering the date of your birthday

d Short-term and long-term memory work separately in the brain. Describe some evidence for this.

_____

_____

_____

12 Scientists have models of how memory works, but they do not explain memory completely.

a Explain, in terms of neuron pathways in the brain, why repetition will help you remember something.

....

....

b Explain, in terms of short-term and long-term memory stores, why repetition will help you remember something.

....

....

c Look at this list of numbers for 20 seconds. Cover the list, then try to write the sequence down.

   1   **2**   **3**   **5**   **8**   **13**   **21**   **34**   **55**   **89**   **144**

....

d If you found it difficult, look for a pattern in the numbers then try again.

....

e Some memories are triggered by particular sounds, sights, or smells. Describe, with an example, how a strong stimulus can make you more likely to remember something.

....

....

....

f Give an example of a method you use to help you remember important information. Then try to explain why it works by using a memory store model.

....

....

....

....

This page is blank

Biology across the ecosystem

1 All organisms are dependent on energy from the Sun.

a Highlight or <u>underline</u> one word to describe the process of transferring energy from the Sun to chemical energy in organic compounds.

autotrophy	photosynthesis	synthesis	chlorophyll	enzyme

b In the diagram representing an ecosystem:

➡ colour **green** the box representing **autotrophs**

➡ colour **red** the box representing the **heterotrophs**

➡ add arrows to show the **transfer of energy** in this system

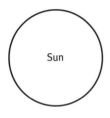

Sun

energy
lost as
heat

producers
(plants)

consumer
(feed off
other living
things)

c Give some examples of autotrophs and heterotrophs in these ecosystems.

Ecosystem	Examples of autotrophs	Examples of heterotrophs
broadleaved woodland		
motorway verge		
garden pond		
ocean		

d Explain how heterotrophs are dependent on energy from the Sun for their food.

....

....

e Some unusual bacteria can obtain energy from raw materials present in the environment. Highlight or <u>underline</u> the correct description of this type of bacteria.

autotrophy	heterotroph

2   Chloroplasts contain the green pigment chlorophyll and the enzymes that are needed for photosynthesis.

a Draw lines to match these key words with their meanings.

carbohydrates	a carbohydrate, synthesized by plants using energy from light
chloroplasts	a carbohydrate, used by plants to store energy
chlorophyll	chemicals made of carbon, hydrogen, and oxygen
glucose	plant cell organelles where photosynthesis takes place
starch	the green pigment needed for photosynthesis

b Complete the word equation that sums up the process of photosynthesis.

$$6CO_2 \quad + \quad 6H_2O \quad \xrightarrow{\text{light energy}} \quad C_6H_{12}O_6 \quad + \quad 6O_2$$

_____ + _____ $\longrightarrow$ _____ + _____

c Answer these questions about the main stages in photosynthesis.

i   What chemical absorbs light energy?

....

ii How is the light energy used?

....

iii What sugar is produced?

....

iv What chemical is produced as waste?

....

d Only a small amount of the sunlight reaching a leaf is used for photosynthesis. List two things that happen to the rest of the light energy reaching a leaf.

1 ....   **2** ....

e Explain why nearly all plants look green.

....

....

3 Glucose made during photosynthesis is used by plant cells in three ways.

a Make notes for each heading to explain how plant cells use glucose. The words in the list should help.

cellulose	chlorophyll	energy	fat	protein	respiration	starch

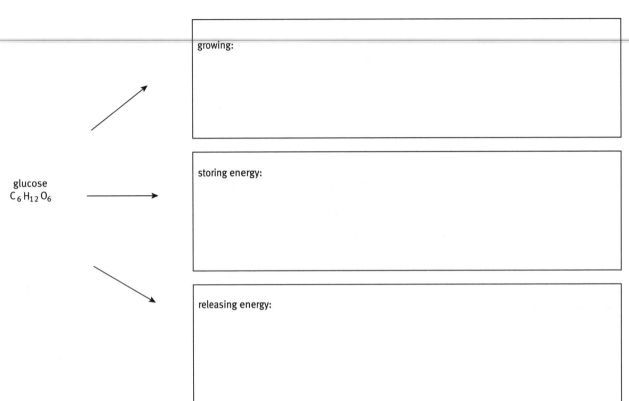

glucose
$C_6H_{12}O_6$

growing:

storing energy:

releasing energy:

b A polymer has long molecules made up of chains of smaller units.
Write down the names of three polymers found in plant cells. (They can be found in the list above.)

1 ....   **2** ....

3 ....

c Energy from cell respiration is used to synthesize polymers.

➔ Complete the diagram showing stages in the synthesis of plant cell polymers.

➔ Colour the arrows that use energy from respiration.

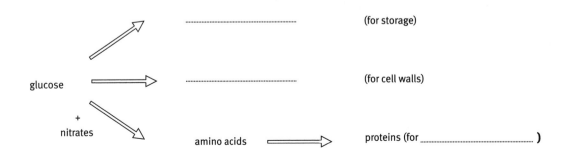

glucose
+
nitrates

.... (for storage)

.... (for cell walls)

amino acids ⟹ proteins (for ....)

d Water can move across cell membranes by osmosis.
Add arrows to show how water would move between the cells shown.

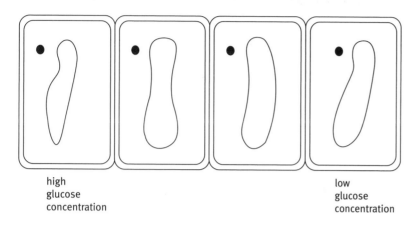

high
glucose
concentration

low
glucose
concentration

e Large carbohydrate molecules such as starch are insoluble. Explain why, for plant cells, starch is a better storage molecule than glucose.

....

....

f Use these words to complete the sentences.

active transport	diffusion	energy	protein

The nitrate ions needed for .... synthesis are taken in from the soil by plant roots.

To move nitrate ions into the cell, against the natural .... gradient, uses

.... This is an example of ....

g Complete the diagram to show the transport of nitrate ions into plant root cells.

➜ Add arrows to show the active transport of the nitrate ions.

➜ Add stars to show where energy is used in this process.

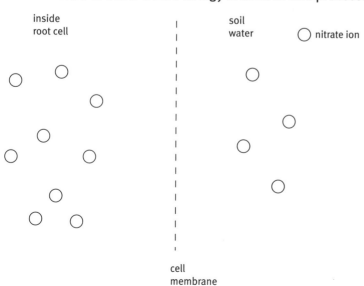

inside
root cell

soil
water   ◯ nitrate ion

cell
membrane

4 The rate of photosynthesis may be limited by several factors.

a Explain how the rate of photosynthesis will affect the growth of a plant.

....

....

b Look at the equation for photosynthesis (see question **2b**) and list four factors that could affect the rate of the process in a healthy plant.
(**Remember:** photosynthesis involves chemical reactions that are catalysed by enzymes.)

1 .... **2** ....

3 .... **4** ....

c Carbon dioxide forms 0.04% of air. The graph shows the effect of light intensity on the rate of photosynthesis in normal air at 20 °C. Use the graph to answer the questions.

i What is the main factor limiting photosynthesis at light intensity A?

....

ii What factors might be limiting photosynthesis at light intensity B?

....

iii Add and label a second line to the graph to show the effect of an increase in the carbon dioxide concentration to 0.1%.

iv Add and label a third line to the graph to show the effect of 0.1% carbon dioxide and an increase in temperature to 30 °C.

d A student collected data about the rate of photosynthesis in pondweed by counting the rate of bubble production under different conditions. Suggest some limitations of the student's data.

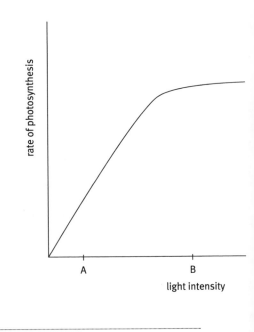

....

....

....

5   Some of the glucose produced during photosynthesis is used during respiration.

a Complete the word equation that sums up the process of respiration.

$$C_6H_{12}O_6 \quad + \quad 6O_2 \quad \longrightarrow \quad 6CO_2 \quad + \quad 6H_2O \quad + \quad energy$$

_____ + _____ $\longrightarrow$ _____ + _____ + _____

b The rates of photosynthesis and respiration in a plant were measured over 24 hours.
Use the information in the graph to answer the questions.

i Describe how photosynthesis and respiration vary over 24 hours.

....

....

....

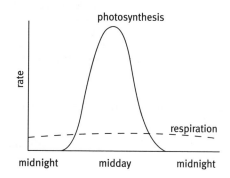

ii Fill in the table.

Chemical	Time of day showing net production	Time of day showing net use
oxygen		
carbon dioxide		
glucose		

iii The compensation point is when photosynthesis and respiration balance so that glucose is produced and used at the same rate. Label two compensation points on the graph.

iv What times of day do the compensation points occur? ....

c Most scientists agree that human activity is causing an increased level of atmospheric carbon dioxide. Fill in the table and then highlight or <u>underline</u> factors that are affected by human activity.

Factors that add CO_2 to the atmosphere	Factors that take CO_2 out of the atmosphere

6 Energy is transferred between organisms in an ecosystem.

a i Choose a type of ecosystem (such as field, woodland, pond, sea, or your own choice) and in the table with examples of organisms that would be found at different trophic levels in that ecosystem.

Ecosystem:	Trophic level	Examples of organisms
	producers	
	primary consumers	
	secondary consumers	
	tertiary consumers	
	decomposers	

ii Use some of these examples to construct a food chain.

producer   → primary consumer   → secondary consumer   → tertiary consumer

.... → .... → .... → ....

iii Explain two ways that energy is transferred between organisms in this ecosystem.

1 ....

2 ....

b i The table shows data collected in a woodland ecosystem.

Woodland	Number of organisms	Biomass (mass units)
producers	5000	54 000
primary consumers	43 000	10 000
secondary consumers	4000	3000
tertiary consumers	1000	500

ii Sketch a pyramid of numbers for this woodland ecosystem.

iii Sketch a pyramid of biomass for this woodland ecosystem.

iv Explain what this data shows about feeding relationships in the food chain.

....

....

c Compare the two pyramids by filling in the table.

Comparisons	Pyramid of numbers	Pyramid of biomass
How easy is it to collect data?		
Does it give a good comparison of the trophic levels?		
What are the main advantages?		
What are some of the limitations		

d i The diagram represents the annual productivity of a sheep farming system. Add arrows to show the energy flow through the system.

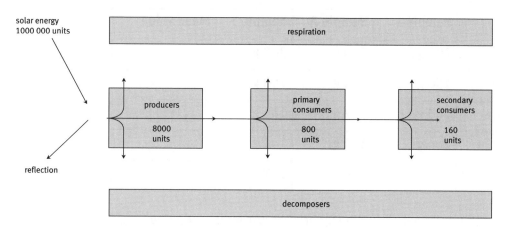

ii Where is the greatest loss of energy in the system?

....

iii If the producers in this system are grasses, the primary consumers sheep and the secondary consumers man, calculate:

⋗ the percentage efficiency of energy transfer from solar energy to grass

$$\frac{\text{energy of grass}}{\text{solar energy}} \times 100 = \frac{\text{........................}}{\text{........................}} \times 100 = \text{............}\%$$

⋗ the percentage efficiency of energy transfer from grass to sheep ....

⋗ the percentage efficiency of energy transfer from grass to man ....

iv This calculation has assumed that all the primary consumers are sheep. What other primary consumers might there be and how would they affect the energy efficiency calculations?

....

7 Soil is an important part of many ecosystems.

a Explain the importance, for plants, of the four main components of soil.

➔ inorganic particles (sand, silt, and clay) _____

➔ water (with dissolved mineral ions) _____

➔ air _____

➔ biomass (living and dead organisms) _____

b Use the data in the table (from several samples of the same soil) to answer the questions.

	Sample 1	Sample 2	Sample 3	Mean value
mass of fresh soil	200 g	200 g	200 g	
mass of dried soil	156 g	154 g	158 g	
mass after heating 5 min	148 g	152 g	144 g	
mass after heating 10 min	144 g	144 g	141 g	
mass after heating 15 min	144 g	146 g	139 g	

i Are there any outliers in the data that should be ignored or further investigated? _____
Explain your answer.

_____

ii Record the mean value for each measurement in the final column of the table.
(Use these values for the following calculations.)

iii Calculate the percentage water in the soil sample.

$$\% \text{ water} = \frac{\text{mass of fresh soil} - \text{mass of dry soil}}{\text{mass of fresh soil}} \times 100 = \frac{\rule{3cm}{0.4pt}}{\rule{3cm}{0.4pt}} \times 100 = \rule{1cm}{0.4pt} \%$$

iv Calculate the biomass as a percentage of the dry soil.

_____

c Explain why the soil sample analysed in **b** could be described as a good fertile soil.

_____

_____

8 Heterotrophs have evolved a variety of feeding relationships.

a Complete the table and add several more examples.

Symbiotic relationship	The word that describes it	Examples
both organisms benefit	mutualism	sea anemone on the shell of a crab
	commensalism	moss on a tree
one organism gains at the expense of another organism		human tapeworm

b Describe the relationship between a host and a parasite.

....

....

c Explain why each of the features listed helps to make the tapeworm a successful intestinal parasite.

  i A tapeworm has male and female sex organs. ....

....

  ii A tapeworm produces a large number of eggs. ....

....

  iii A tapeworm head has suckers and hooks. ....

....

  iv A tapeworm has a large surface area. ....

....

  v A tapeworm can use anaerobic respiration. ....

....

d Add notes to the diagram to explain the features that enable the malaria parasite to be successful.

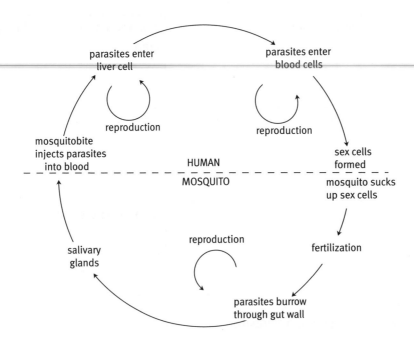

e Complete the table to describe some of the effects parasites can have on farm animals and crops.

Parasite	Impact on food production
roundworms in cattle and sheep	
potato blight (a fungus)	
fish lice on farmed salmon	
liver fluke in dairy cattle	
canker (a fungus) on apple trees	

f Explain why the evolution of a parasite is thought to be closely linked to that of its host.

....

....

....

9 Natural selection has resulted in an increased frequency of the sickle-cell allele in certain populations.

a Sickle-cell anaemia is caused by an allele of the gene which codes for haemoglobin. The sickle-cell allele is recessive. Use two colours and a key to show:

➔ members of this family who have sickle-cell anaemia

➔ members of this family who are carriers and have the sickle-cell trait (one normal allele and one sickle-cell allele)

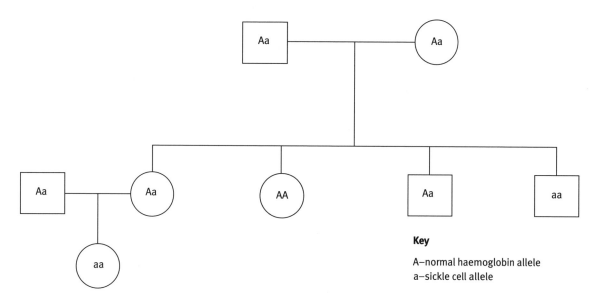

Key
A—normal haemoglobin allele
a—sickle cell allele

b What symptoms would someone who inherited two sickle-cell alleles show?

....

....

c Carriers of the sickle-cell allele have some protection from malaria.

i Explain how natural selection has led to the sickle-cell allele being rare in parts of the world where the malaria mosquito cannot survive.

....

....

....

ii Explain how natural selection has led to the sickle-cell allele being more common in parts of the world where malaria is a common cause of infant mortality.

....

....

10 Bacteria and fungi can be grown on a large scale to give useful products.

a Draw lines to match each key word with its meaning.

antibiotic	a drug that stops the growth of, or kills, bacteria and fungi
enzyme	food protein obtained by growing microorganisms
fermentation	a group of single-celled fungi
single-cell protein	growing bacteria or fungi to make a useful product
yeasts	a protein that catalyses a chemical reaction

b Draw a bacterium and label the structures listed.

cell wall

cell membrane

circular DNA chromosome

DNA plasmid

c Give examples of products of traditional technologies:

➔ using yeast fermentation _____

➔ using an enzyme extract _____

d New technologies enable a wide range of products to be made by fermentation of carefully selected bacteria and fungi. Complete the table.

Type of product	Examples
antibiotics	
	Quorn, animal feed proteins
	rennin, amylase, pectinase, glucose isomerase

11 Genetically modified (GM) organisms are used to make new products or improve efficiency.

a In the diagram showing the genetic modification of a bacterium, add notes to describe the three main steps in the process.

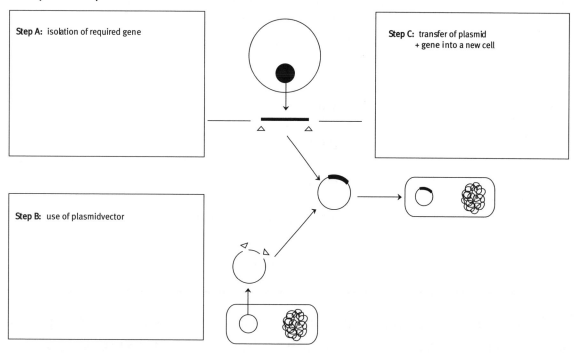

Step A: isolation of required gene

Step C: transfer of plasmid + gene into a new cell

Step B: use of plasmidvector

b Use these words to complete the sentences.

bacteriophage	plasmid	vectors

In the above example of genetic modification of a bacterium, the vector is a _____ .

A virus, such as a _____ , is sometimes used as the vector during the genetic

modification of bacteria. Viruses and bacteria are used as _____ in the genetic

modification of plants.

c Outline the main steps necessary to transfer gene X from a bacterium to sugar beet using a virus vector. The gene X makes the sugar beet more disease-resistant.

➡ **Step A** _____

_____

➡ **Step B** _____

_____

➡ **Step C** _____

_____

d Use two different colours to highlight the types of organism used in fermentations (box 1).
Then use the colours to match them up with the useful products of fermentations (box 2).
(At least one product is made two alternative ways, so could be coloured with two different colours.)

1

GM fungi or bacteria
natural fungi or bacteria

2

penicillin
insulin
alcohol
Quorn
rennin
human growth hormone

---

12 There are economic, social, and ethical implications for the release of GM organisms.

a The potential risks of introducing GM organisms into the environment need to be balanced against the potential benefits. Use the table to summarize some of the arguments about the examples given.

	Arguments in favour	Arguments against
Question 1: Should a company be allowed to make a new antibiotic against tuberculosis in a fermentation system using GM bacteria?		
economic		
ethical		
social		
Question 2: Should farmers be allowed to grow insect-resistant maize from GM seeds?		
economic		
ethical		
social		

b Read about some of the implications of growing insect-resistant maize then answer the questions below.

> A variety of maize has been genetically modified to be resistant to the European Corn Borer. This insect pest bores into maize plants causing them to fall over. In infested regions it can destroy 20% of the crop.
>
> The pest is traditionally controlled using insecticide sprays. These work only during the first three days in the corn borer's life cycle, before it bores into the plant stem. The new variety of GM maize produces Bt toxin, which kills the corn borer. The Bt gene comes from a bacterium which is used as a biological insecticide by organic farmers.

Why is the introduction of any new crop variety, developed by cross breeding or genetic modification, never completely safe?

What sort of regulations and laws are involved?

Suggest how the risks of growing the maize could be reduced.

Can using these seeds be justified or is it unnatural or wrong?

use of GM organism

Explain why the 'precautionary principle' would oppose growing the maize.

What is the best outcome for the majority of people involved?

13 DNA can be extracted from cells for genetic tests.

a Genetic tests have been devised to provide information about a person's DNA.
List some of the ways that information from genetic tests is used.

b DNA for a test is extracted from cells.

i Draw a simple diagram of a white blood cell, and label the features listed.

nucleus

cytoplasm

cell membrane

nuclear membrane

ii Where is the DNA in the cell?

iii What membranes must be opened to extract the DNA?

iv The DNA is then purified. What substances need to be removed from the cell extract?

v Why are red blood cells not used for DNA testing?

14 DNA technology is used to prepare gene probes for each test.

a A gene probe is a short piece of single-stranded DNA which has complementary bases
to a DNA chain of the test allele. Add the correct bases to make a probe for the length of
DNA shown.

marker
gene probe

T T A C G G T G C A A T C

test
allele

Remember : C pairs with G, T pairs with A

b Use these words to complete the description of how a gene probe works.

complementary	DNA	marker	stick

Genetic tests that use probes rely on the fact that when _____ is gently heated

and cooled the double strands separate and then rejoin. If this is done in the presence

of a gene probe, the probe will _____ to sequences of DNA in the test sample that

are _____ to the probe. The _____ on the probe shows if this

has happened.

c Markers for gene probes can be fluorescent or radioactive. Explain how the probes are detected in each case.

➔ Fluorescent probes: _____

_____

➔ Radioactive probes: _____

_____

d Explain the meaning of each of these terms used in DNA technology:

➔ gene _____

➔ allele _____

➔ gene probe _____

➔ autoradiography _____

➔ UV _____

15 Blood tissue has several different components.

a Match each blood component with the information about it using lines or colours.

Number of cells in 1 mm³ of blood	Blood components	Function
7000	red blood cells	to transport water, solutes, and heat
250 000	white blood cells	to clot blood at injury sites
5 000 000	platelets	to fight infection
no cells (liquid)	plasma	to transport oxygen

b Draw the following cells, and label with notes to explain how they are adapted for their function:

a white blood cell   a red blood cell

c Draw lines to match these key words to their meanings.

antibody	bloods that can be mixed without antigen–antibody problems
antigen	a marker on a cell surface that can cause antibody formation
blood transfusion	a person who receives blood from another person
compatible bloods	a protein that recognizes and binds to particular antigens
donor	the process of transferring blood from one person to another
recipient	a person who gives blood to another person

d Every person has one of four ABO blood types. The A, B, AB, or O blood types describe what ABO antigens are present on the red blood cells (A, B, both, or neither). There are also antibodies present in the blood plasma (anti-A and/or anti-B). Each blood type has antibodies against the ABO antigens that are not present on the red cells. Complete the table to show this.

Blood type	Antigens on the red cells	Antibodies in the plasma
A		
B		
AB	and	none
O	neither	and

e Blood is an important tissue and needs replacement after loss through accident or surgery. Blood transfusions are only possible between certain ABO blood types. The recipient must not have plasma antibodies against the red cell antigens in the donated blood.
Fill in the table:

➔ list the recipient antibodies present for each blood type

➔ add ticks ✓ or crosses ✗ to show if transfusion is possible for each donor/recipient combination

Recipient		Donor type A	Donor type B	Donor type AB	Donor type O
blood type	antibodies				
A					
B					
AB					
O					

f What blood type is the 'universal donor'? Explain what this means.

....

....

g What blood type is the 'universal recipient'? Explain what this means.

....

....

16 Your ABO blood type is determined by a single gene with three alleles.

a The three ABO alleles are I^A, I^B and I^O. The alleles I^A and I^B are codominant. The allele I^O is recessive to both. Complete the table to show how different pairs of alleles lead to the four ABO blood types.

Allele pair	Blood type
I^A I^A	
I^A I^O	
I^B I^B	B
	AB
	O

b Predict the possible blood types of children in the following families.

Family 1

	Father type A alleles I^A I^O	
Mother type A alleles I^A I^O	alleles	alleles
	blood type	blood type
	alleles	alleles
	blood type	blood type

Family 2

	Father type O alleles I^O I^O	
Mother type AB alleles I^A I^B	alleles	alleles
	blood type	blood type
	alleles	alleles
	blood type	blood type

c What chance is there of a child being blood type A in each of the above families?

⮕ Family 1 ....

⮕ Family 2 ....

d What ABO antibodies would be present in the blood plasma of the mother and father of family 2?

⮕ Mother ....

⮕ Father ....

17 The heart pumps blood around the body.

a Complete the table describing different parts of the heart.

Heart structure	Description
	receives blood from all parts of the body except the lungs
	pumps blood to the lungs
	receives blood from the lungs
	pumps blood to all parts of the body except the lungs
heart valves	

b In the above table:
- ➜ colour **red** the heart structures that are filled with **oxygenated** blood
- ➜ colour **blue** the heart structures that are filled with **deoxygenated** blood

c Complete these sentences.

Blood is carried towards the heart by blood vessels called ....

Blood is carried away from the heart by blood vessels called ....

d Valves are found in the heart and in veins.
- ➜ Draw a ring around the positions of the four heart valves.
- ➜ Explain what the valves in veins do.

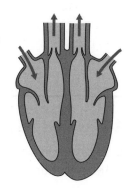

....

....

e Complete the diagram to show how the valves in veins work.

direction
of
flow

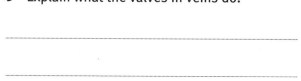 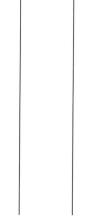

valve open
allowing flow

valve closed
preventing back flow

g In the flow chart:

⮆ add arrows to show the double circulation of the blood around the body

⮆ label the points where there is a capillary network

lungs

heart

rest of body

h Explain why a double circulation is important for oxygenated blood to reach all parts of the body.

....

....

---

18 Aerobic respiration provides most of the energy to your body cells.

a Write the word equation that summarizes the process of aerobic respiration, then answer the questions.

.... + ... → .... + ... + energy

i What is the body's source of glucose for respiration? ....

ii What is the body's source of oxygen for respiration? ....

iii How is the carbon dioxide waste removed from the body? ....

iv How is the glucose, oxygen, and carbon dioxide transported around the body? ....

....

b The exchange of chemicals between the blood and body cells takes place in the capillary beds and tissue fluids. Answer the questions about this process.

i How is the tissue fluid formed?

ii By what process are chemicals exchanged between the blood and body cells?

iii Name two chemicals passing from the blood into body cells.

iv Name a waste product from aerobic respiration passing from body cells into the blood.

v Name a waste product from protein and amino acid breakdown passing from liver cells into the blood (before being excreted by the kidneys).

c During exercise muscle cells need more energy. Explain how the following changes help to provide the extra energy needed by the muscle cells.

⮑ The breathing rate increases.

⮑ The heart rate increases.

d Explain why 'normal' measurements for heart rate and blood pressure cover a range of values.

19 Quick bursts of energy can be provided by anaerobic respiration.

a Anaerobic respiration can help to provide energy by using glucose but not oxygen.
Write the word equation that summarizes anaerobic respiration.

.... → .... + energy

b Explain why anaerobic respiration can only be used by muscles for a short period of time.

....

....

c Oxygen is then needed to break down the lactic acid. Explain why this is called 'oxygen debt'.

....

....

d Summarize the differences between aerobic and anaerobic respiration by filling in the table.

	Aerobic respiration	Anaerobic respiration
What is the energy source?		
What else is needed?		
What waste products are formed?		
Compare the efficiency (which produces the most energy per molecule of glucose).		
Give some examples of when it is useful to the body.		
Describe the effects of increasing this respiration.		
Describe any after effects of increased respiration.		
Other notes		

20 For body movement, muscles need a supply of energy (as ATP) and a supporting framework.

a Use these words to complete the sentences.

ATP	currency	glucose	released	stored

All respiration releases energy from .... The energy is ....

in the chemical ATP. Cell processes that require energy break down .... ,

energy is then .... ATP can be described as the energy ....

of living things.

b A muscle is made up of hundreds of muscle fibres. Each muscle fibre is packed with protein filaments. Describe what happens to the fibres when a muscle is used to move a part of the body.

....

....

c Complete the flow chart to show the transfer of energy from glucose in the blood to the cells in a contracting muscle. Label the processes involved (include both aerobic and anaerobic respiration).

| food energy as blood glucose |

| energy used in muscle contraction |

d Draw lines to match these keys words to their meanings.

external skeleton	a living structure that provides a jointed framework to support movement, typical of vertebrates
internal skeleton	an animal with a backbone
vertebrate	a hard outer covering providing support and protection, may have flexible joints
invertebrate	an animal that does not have a backbone

e Give two examples of:

➔ invertebrates with an external skeleton

1 .... 2 ....

➔ vertebrates with an internal skeleton

1 .... 2 ....

f Describe some features of the human skeleton that help it to:

i provide support ....

....

ii protect the brain and spinal cord ....

....

iii allow movement ....

....

iv provide new blood cells ....

....

v store minerals ....

....

21 Movement of a joint depends on all its parts functioning effectively.

a i Draw two muscles that move the elbow joint on the diagram of the bones of the arm.
Label the muscles A and B.

ii Explain what happens to each muscle when the lower arm is raised.

A .... **B** ....

iii Explain what happens to each muscle when the lower arm is lowered.

A .... **B** ....

iv These muscles are an antagonistic pair. Explain what this means.

....

....

v Describe an antagonistic pair of muscles from another part of the body.

....

b Draw lines to match these keys words to their meanings.

cartilage	fluid lubricating and nourishing a joint
joint	tissue joining bones together
ligament	tissue joining muscle to bone
synovial fluid	smooth, shock-absorbing tissue protecting bones
tendon	where two or more bones meet

c Draw and label a joint between two bones showing: bones, ligaments, cartilage, synovial fluid.

d Explain the function of the synovial fluid.

....

e Specific properties of joint tissues enable them to function effectively. Complete the table:

➔ describe the function of each tissue

➔ use words from this list to describe some specific properties of each tissue

elastic (bouncy)	fibrous	flexible	not elastic
not stretchy	pearly	smooth	strong	white

Tissue	Function	Specific properties
cartilage		
ligaments		
tendons		

22 The correct treatment of skeletal-muscular injuries can help the healing process.

a Identify these common injuries from their descriptions.

| dislocation | sprain | torn ligament | torn tendon |

Description	Injury
tissue damage that results in an unstable joint	
tissue damage that results in loss of a certain movement	
an overstretched ligament	
displacement of a bone from its normal position in a joint	

b The most common sporting injury is a sprain. Complete the descriptions of the three main symptoms of a sprain.

1 Appearance (shape): ....

2 Appearance (colour): ....

3 ....

c The immediate treatment for sprains is RICE. Explain what each letter stands for, and describe what you would do to help someone who had sprained an ankle.

....

....

....

....

....

d Describe the sort of treatment that might follow a suitable period of RICE treatment.

....

....

e Describe the role of the physiotherapist in treatment of skeletal-muscle injuries. At what stage would a physiotherapist be involved? What sort of specialist help would a physiotherapist give to aid full recovery?

....

....

....

....

f Read the list of exercises recommended to someone recovering from a joint injury, then answer the questions below. Include references to muscles, tendons, and ligaments in your answers.

Exercise 1
Lie with your leg out straight. Tense up your thigh muscles, push your knee down, and try to raise your heel. Hold for a few seconds.

Exercise 2
Place a rolled up towel under your knee, keep your knee on the roll, and lift your heel. Try to get your knee completely straight.

Exercise 3
Bend your knee as far as it can easily go. Hold for a few seconds then straighten and repeat.

Exercise 4
Lie on your front. Keep your thigh down and bend your knee as far as you easily can.

---

i What do you think is the main purpose of these exercises?

....

ii Why do you think all the exercises are done lying down?

....

iii What would be the aim of exercises in the next stage of the recovery process? What sort of exercise might be recommended?

....

iv What would be the aim of exercises recommended for a final return to full fitness?

....

23 Physical training programmes can aid recovery from illness or injury, or improve fitness.

a Make a list of the sort of medical and lifestyle information that a trainer or medical professional would need to know about a client before recommending a training programme.

b Explain how a trainer working with an athlete would use their medical and lifestyle history.

c Explain how a hospital physiotherapist treating someone recovering from a joint operation would use their medical and lifestyle history.

24 Accurate record keeping during treatment or fitness training is essential.

a Health or fitness practitioners like to have regular contact with their patients or clients. Explain the advantages of this for:

→ the health or fitness practitioner

→ the patient or client

b Why does a fitness trainer need to know if their client starts a new course of medication?

c A good health or fitness practitioner team ensures 'continuity of care'.

i Explain the importance of accurate record keeping.

....

ii Explain the importance of the way those records are stored.

....

iii Explain the importance of allowing the right professionals access to the records.

....

d Describe an example of how the progress of a sportsman or sportswoman is monitored over a season of training. Give examples of what measurements would be taken and recorded.

....

....

....

....

---

25 Good training programmes include assessment, modification, and follow up.

a Treatments often have side effects which have to be weighed up against the benefits. Add examples of possible benefits and side effects for the treatments suggested in the table.

Suggested treatment	Possible benefits	Possible side effects
A a gym workout programme for an overweight person (Target: weight in the 'normal' range)		
B a walking programme for a patient with heart problems (Target: improved cardiovascular fitness)		
C a very strict, intensive training schedule for an athlete (Target: optimum fitness for a particular event)		

b Choose two of the examples (A, B or C) in the table, and for each suggest why the programme might need to be modified before it has been completed.

⇨ Example .... : ....

....

⇨ Example .... : ....

....

c Choose two of the examples in the table, and for each suggest another way of achieving the target.

⇨ Example .... : ....

....

⇨ Example .... : ....

....

d If the patients/clients in the table achieved their targets, suggest how their progress could then be monitored.

A ....

B ....

C ....

e Assessment of progress needs to take into account the reliability of the data obtained. Explain what problems there might be with data obtained in these situations.

⇨ **Example A** (in the table) occasionally measuring his/her weight at home using bathroom scales.

....

⇨ **Example B** having blood pressure and pulse measured weekly by a visiting nurse who calls at a different time each week.

....

⇨ **Example C** doing regular timed runs on an outside track in different weather conditions.

....

TWENTY FIRST CENTURY

science

GCSE Biology

Nuffield
Curriculum Centre

OCR
RECOGNISING ACHIEVEMENT

THE UNIVERSITY *of York*

OXFORD

The exercises in this Workbook cover the OCR requirements for each module. If you do them during the course, your completed Workbook will help you revise for exams.

Project Directors
Jenifer Burden
John Holman
Andrew Hunt
Robin Millar

Course Editors
Jenifer Burden
Peter Campbell
Andrew Hunt
Robin Millar

Project Officers
Peter Campbell
Angela Hall
John Lazonby
Peter Nicolson

Authors
Jenifer Burden
Caroline Shearer

Contents

WORKBOOK

UNIVERSITY PRESS

Great Clarendon Street, Oxford OX2 6DP

Oxford University Press is a department of the University of Oxford.
It furthers the University's objective of excellence in research, scholarship,
and education by publishing worldwide in

Oxford   New York

Auckland   Cape Town   Dar es Salaam   Hong Kong   Karachi
Kuala Lumpur   Madrid   Melbourne   Mexico City   Nairobi
New Delhi   Shanghai   Taipei   Toronto

With offices in

Argentina   Austria   Brazil   Chile   Czech Republic   France   Greece
Guatemala   Hungary   Italy   Japan   Poland   Portugal   Singapore
South Korea   Switzerland   Thailand   Turkey   Ukraine   Vietnam

British Library Cataloguing in Publication Data

Data available

ISBN: 978-0-19-915052-6

10 9 8 7 6 5 4 3

Printed in Spain by Unigraf

Illustrations by IFA Design, Plymouth, UK

These resources have been developed to support teachers and students undertaking a new OCR suite of GCSE Science specifications, Twenty First Century Science.

Many people from schools, colleges, universities, industry, and the professions have contributed to the production of these resources. The feedback from over 75 Pilot Centres was invaluable. It led to significant changes to the course specifications, and to the supporting resources for teaching and learning.

The University of York Science Education Group (UYSEG) and Nuffield Curriculum Centre worked in partnership with an OCR team led by Mary Whitehouse, Elizabeth Herbert and Emily Clare to create the specifications, which have their origins in the Beyond 2000 report (Millar & Osborne, 1998) and subsequent Key Stage 4 development work undertaken by UYSEG and the Nuffield Curriculum Centre for QCA. Bryan Milner and Michael Reiss also contributed to this work, which is reported in: 21st Century Science GCSE Pilot Development: Final Report (UYSEG, March 2002).

Sponsors
The development of Twenty First Century Science was made possible by generous support from:
• The Nuffield Foundation
• The Salters' Institute
• The Wellcome Trust